Frankel-y Speaking About:

WWII in the South Pacific

By Stanley A. Frankel

Library of Congress Cataloging-in-Publication Data

Frankel, Stanley A., Frankel-y Speaking About WWII in the South Pacific, memoir of military service in WWII

Summary: Stanley Frankel, the official historian of the 37th Infantry Division in the Pacific in World War, II tells his own story, of serving alongside Private Rodger Young who gave up his life in New Georgia to save twenty men of his patrol and inspired a song about the rescue of Bilibid Prison, of the battles of Bougainville and Guadalcanal, and of the daily letters he wrote to the sweetheart who would become his wife.

ISBN: 978-1-939282-43-9

Published by Miniver Press, LLC, McLean Virginia
Copyright 2019 Nell Minow

All rights reserved under International and Pan-American Copyright Conventions. By payment of the required fees, you have been granted the non-exclusive, non-transferable right to access and read the text of this e-book on-screen. No part of this text may be reproduced, transmitted, down-loaded, decompiled, reverse engineered, or stored in or introduced into any information storage and retrieval system, in any form or by any means, whether electronic or mechanical, now known or hereinafter invented, without the express written permission of Nell Minow. For information regarding permission, write to editor@miniverpress.com

Second edition December 2019

Table of Contents

Introduction from Adam P. Frankel 7
Introduction from Stanley A. Frankel 9
Acknowledgments 13
Dedication 15
Chapter 1: A Reluctant Soldier 17
Chapter 2: Every Night We Die 23
Chapter 3: The New Champions 33
Chapter 4: The Last Ballad 41
Chapter 5: To The Rescue 49
Chapter 6: Taking it From the Air 55
Chapter 7: Japanese Soldiers, Docile and Cooperative 61
Chapter 8: Once Upon a Christmas 69
Chapter 9: No Rest Areas 73
Chapter 10: Bougainville Mopup, and Ready for Philippines 81
Chapter 11: The Luzon Beachhead 91
Chapter 12: The Battle of Balintawak 97
Chapter 13: The Rescue of Bilibid Prison 103
Chapter 14: Incidents in the Battle for Manila 113
Chapter 15: The Damn Fool 123
Chapter 16: The Baguio Campaign 129
Chapter 17: The End of the War 137
Chapter 18: An Unexpected Assignment 145
Chapter 19: Return of the Heroes 149
Chapter 20: Afterthoughts 155
Appendix 163
    Awards 165
    Birth of a Patriot 181
    Aye or Nay? 187
    Adlai Remembered 193
    A Not-So-Grand-Fathers' Day 197

Wordsmith for Presidential Hopefuls	201
Teaching at Baruch	205
If I Were 21	209
Rudolph that Amazing Reindeer	215
The Tragic Truth About Our Jury System	221
A Baseball Memoir	227
Thoughts on the Fourth, and the Fourteenth, and…	229
Clout, You Gotta Have Clout	233
The Enemy Is Us, Not Saddam	235
Where Have You Gone, Joe DiMaggio?	237
"Pa, Are You Ever Going to Die?"	241
Angry with Senator Quayle, Soldier?	245
Frankel-y Speaking	249

DR. MARC HOLLENDER, MY EX-ROOMMATE AND
LONGTIME FRIEND, KEPT ALL THE LETTERS I
WROTE HIM DURING WWII. RECENTLY, HE SENT
ME A PACKAGE OF 53 OF THEM...AND THIS ONE
IS, I THINK, ONE OF THE MORE AMUSING.

27 May, 1945
Philippines

Dear Marc,

      Your book, People On Our Side, just arrived, and I am afraid it is going to interfere with the war effort..at least my own contribution which, between us, doesn't make a helluva lot of difference anyway.

      For the past six months we have been driving so hard that I just got out of the book habit. I carried my pocket edition of Time and New Yorker in my pocket, hastily read a story in between air raids, chow, or early morning plumbing, and carried on. In fact, now that we have become a bit more settled, I just haven't had the patience to stay with any one book. However, this one by Snow really sounded good, and I started in on it during a lull in my paper work..and had one hard time putting it down. Thanks ever so much, both for your thoughtfulness and for your good taste.

      I have received letters from you while on the move, but we are not permitted to hang onto these letters...security..in case we are found dead with them...so I burn yours, memorize the address..and promptly forget it..necessitating my writing through Irene. This time, I think I have that address licked..and if I send this one through Irene, you'll know the address was lost during the last five minutes before typing it on the envelope.

      We have engaged in very strenuous operations, and even though I am not digging out the Nips with my bayonet, I have had my moments..which I would have sold out my chances for survival for 10% pre-war prices. While ducking a truckload of Nip artillery, mortar, and rocket shells at Malacanan Palace in Feb., I tried to figure out a way to get into the Aid Station..the one really safe place in the Palace..surrounded by double walls..and in a bomb proof shelter. So, seeing a couple of wounded men lying around the area, I figured that they were my ticket into the aid station. Thought I:..carry one with you and they'll have to let you in. Once in, let them try to get you out..even if you have to stretch out with the corpses over on the right. So, selfishly, I made the 100 yards to the wounded in four seconds, looked around for the lightest guy and found a semi-midget, picked him up, got to the aid station, was practically dragged out again by a three man litter team who needed a fo'th...so I volunteered..with one guy on each of my arms and the third with a gun in my back...and thus I made a couple of more trips...which caused me extreme nervousness and exhaustion... and the upshot of it all is that my Regtl. Commander, watching from afar,.so he couldn't read my expression, put me up for a Silver Star. Doubt if I'll get it because the Awards Board at higher Hq. knows that I am not the hero type and something is fishy...but it just shows how ironic a guy's instinct for self-preservation results in.

And if you think I am fooling in the last paragraph you are dead wrong. I did not jump; I was pushed.

Anyway, maybe this new discharge plan will affect me someway. I have 98 points, will make 103 if I can get this award; that's a lot most places, but in my regiment with men who have been wounded five times (25 points); with just as much time overseas, in the army, and number of campaigns..plus a DSC or a SS tossed in...well...I rank only in the upper third of the officers ; and I don't visualize "Santa Claus" Marshall breaking up this team just because we happen to have the required number of points. And Marc, this is a real team...with five good fights under our belts, way over 100% casualties and almost 100% medals for heroism; several companies cited by the President, and the Regiment up for one for Luzon; that's a tough combination. The Regiment has killed 18,657 Japs these last three years, and they really smell bad. Which is about my major contribution...since my platoon leading days I am no longer an eager beaver.

Nothing else, Marc, hope this war-mind of mine hasn't bored you... but it's tough to live something for three years without it becoming a part of you.. for better or worse.

Thanks again,

---

* Transcription of letter in the appendix

Introduction from Adam P. Frankel

WHEN I WAS A CHILD, I idolized my grandfather Stanley Frankel—Pa, as I always called him. A big part of the reason is what's in this book—stories of his experiences in the South Pacific during World War II.

I'd curl up next to him in my child-size camouflage uniform as he told me about snipers, flamethrowers, and the heroes he'd fought alongside in the jungles of New Georgia and the streets of Manila.

Like so many members of his generation, Pa often wrote about the war. And I always believed that writing about it—often, immediately after combat ceased—helped him process the trauma of the experience.

In fact, what's now called "expressive writing" is used by veterans' hospitals and organizations as a way of managing post-traumatic stress disorder (PTSD)—what was, in Pa's day, called "battle fatigue" or "shell shock."

Still, Pa never left the trauma of the war completely behind him. When Steven Spielberg's *Saving Private Ryan* came out in 1998, I asked if he wanted to see it together. He declined, explaining that, based on everything he'd read, the film was such a realistic depiction of events that he was concerned about the memories and feelings it would stir up.

Reading Pa's stories illuminates why he was so frightened of those feelings. He puts you right there with him, in the foxhole, on the beach, scared, horrified, by what he's witnessing, what he's doing, what he's experiencing.

This is a gripping, vivid account of a young American at war. But it's also something more. It's a reminder of the great cost at which freedom from fascism and tyranny was purchased by Pa and every other member of his generation.

At a time when leaders in the United States and around the world seem intent on constraining or rolling back the rights, laws, norms, and institutions necessary for a free society to survive and flourish, these stories reveal what happens when we allow extremism, nativism, and visions of racial superiority to take root.

Pa's stories call on us to remember—and to honor those memories by fighting for our ideals and safeguarding our democracies.

>Adam P. Frankel
>Author of *The Survivors: A Story of War, Inheritance, and Healing*
>New York City, 2019

Introduction from Stanley A. Frankel

I WAS DRAFTED into the U.S. Army on January 23, 1941, ten months before our nation entered World War II. I was assigned to the 37th Ohio National Guard Division. A very reluctant soldier, I believed, as did most of those drafted with me, that I would serve no more than twelve months. Five years and five major battles later I was discharged with the rank of major.

Like most American men in my generation I was not brought up to be a soldier. My first twenty-three years ill-prepared me to be plunked down as a World War II infantryman for 3½ years—in the dank jungles of the Solomon Islands, in the lethal streets and buildings of Manila, and on the frightening, winding mountain roads leading up to Baguio in the Philippines. Those who knew me in Dayton, Ohio, where I grew up, and at Northwestern University where I led student rallies calling for non-participation in the war in Europe, surely would have thought it wildly unlikely that I would become an officer in the U.S. Army in the South Pacific leading troops into battle.

The chapters in this book are my own personal account of what might be called "Frankel's War." They are based on recollections and materials written at the time. Many of the chapters are accounts in the form of letters drafted immediately after combat ceased, red hot on a borrowed typewriter, and sent to a "Dear Aunt Madeleine" in New York City. Aunt Madeleine was really Ms. Madeleine Brennan, a top literary agent, who had been persuaded by my father-in-law to be, Salem Baskin, to read and, if warranted, try to sell what I had sent her. Some of these she managed to get published; the balance have been in my files for all these years, carbons on onion-skin paper of the originals I had sent to "Aunt Madeleine."

Then there were personal letters. During my entire service, my college sweetheart and almost-fiancée, Irene Baskin, to whom I had given my fraternity pin and Phi Beta Kappa key, wrote me every other day, and I responded faithfully and regularly. These letters to her, exactly 1,232 of them, she kept in loose-leaf notebooks and presented to me on my return

Stanley A. Frankel

from the Pacific in January 1946. (We were married February 20, 1946.) Rereading these letters 45–50 years later called to mind many hundreds of half-forgotten incidents and triggered some of the materials that follow.

Those of you familiar with the ways of the army in World War II may wonder how I managed to send materials like those to "Aunt Madeleine" to the States uncensored. At first I had few problems. All outgoing mail written by enlisted men was censored by their immediate commanding officer (quite an inhibitor to the love-and-action comments of the soldiers). Letters from officers were censored in random fashion by the Army Post Office in the field, the usual hit-or-miss affair resulting in few hits and many misses.

For a long while all my letters got through, untouched. Eventually, however, when some of my stuff started being published in the States in newspapers and national magazines, clippings would often be relayed by the loved ones back home to the commanding general, colonels, and others.

Articles intended for publication were supposed to be submitted to the Division G2, a procedure I had not wanted to risk because of my apprehension that I would be denied permission. When the top brass did get wind of my publications, they did not make an issue with me about my breach of army regulations. But thereafter the Division Post Office examined every Frankel letter carefully. Irene began to receive V-Mail with words and sentences clipped out. One, she recalls, had a sentence beginning: "I am on an...(next word scissored out)." It didn't take her too long to fill in the missing "island." Later, after I found out I was being censored, I resorted to some tricks to slip out information. Once, trying to let her know where I was, I asked her repeatedly about a "mutual friend, Helen."

Irene wrote back, "Which Helen?"

I then replied, "The Helen who was a sorority sister of yours at Northwestern."

She answered, "Do you mean Helen Goldberg or Helen Solomon?" And she further wanted to know why I had suddenly evinced an interest in them, since we had not been particularly friendly with either.

I reacted: "No, not Helen Goldberg. The one I really want you to concentrate on is the other Helen."

At that, Irene figured it out that I was trying to tell her we were in the Solomon Islands.

In addition to the letters to Madeleine Brennan and to Irene, I had my diary, which triggered many memories. Keeping a diary was forbidden, but I had kept one nevertheless, in my back pocket during combat, hidden in my footlocker between missions. This diary served as the source of several of these chapters.

Stanley A. Frankel

Acknowledgments

A NUMBER OF FRIENDS ENCOURAGED...hell...pressured me to dust off the keys of my Royal 400 and bang out this volume. Dr. Marc Hollender, my old and dear buddy, wrote, phoned, and needled me at least once every month for twenty years to put this on paper. My new friend, Dr. Patrick Quinn, Northwestern University archivist, read my papers and urged me to go public with my World War II experiences. My Scarsdale neighbor and old friend, Dr. Rowland Mitchell, edited, rewrote, and guided the direction and the content of this whole manuscript.

And then, of course, there were so many of my compatriots who acted on the often-bloody stage on which we fought. I cannot mention all of their names here but I must list some of those who never returned to read this: Bob Richter, killed in Malacañan Palace in Manila just as I ran out to order him to come inside; Warrant Officer Sayre Shulters, hit by Japanese artillery in the same Manila bombardment; famed Pvt. Rodger Young who gave up his life in New Georgia to save twenty men of his patrol; Lt. Bob Viale, who fell on a hand grenade in Manila to keep it from getting ten soldiers in his platoon; Lt. Bob Harley, hit by a mortar shell in New Georgia while he and I were eating C rations at dusk; Al "Shorty" Gold, fleet center fielder on my regimental baseball team who could not outrun a Japanese mortar shell in Manila; Lt. Bob Hollomon, whom I trained on the Canal with other reinforcements and who was gunned down by an enemy machine gun on the outskirts of Manila...and on and on. These dear, courageous friends and comrades gave that "last, full measure of devotion" not only to their country but to their fellow soldiers and me. I would hope that their wives, families, children will read this book and know they are not forgotten; that they have achieved a kind of immortality in these short and insufficient lines. I only wish that these words would never have had to be inscribed.

Stanley A. Frankel

At war's end I was assigned to write the history of the 37th Division using as sources the records written by hundreds of officers and men of the division. A 400-page volume, it recounts rather straightforwardly the division's experiences in training, in combat, and in the long interludes between battles. Although I am listed as author, the 40,000 men who served in the division really wrote that history.

This more personal account is mine alone.

Dedication

I DEDICATE THIS EFFORT to Irene, who suffered through those difficult war years, and who has given me the love, encouragement, and devotion which have enriched and fulfilled my own life. To my three beloved children, Stephen, Thomas, and Nancy, and her husband, Henry Joselson, our new son, this book is also dedicated. And, finally, it is dedicated to my grandson Adam, in the hope and prayer that what happened to me and my generation will never, never, never happen to Adam and his.

Stanley A. Frankel

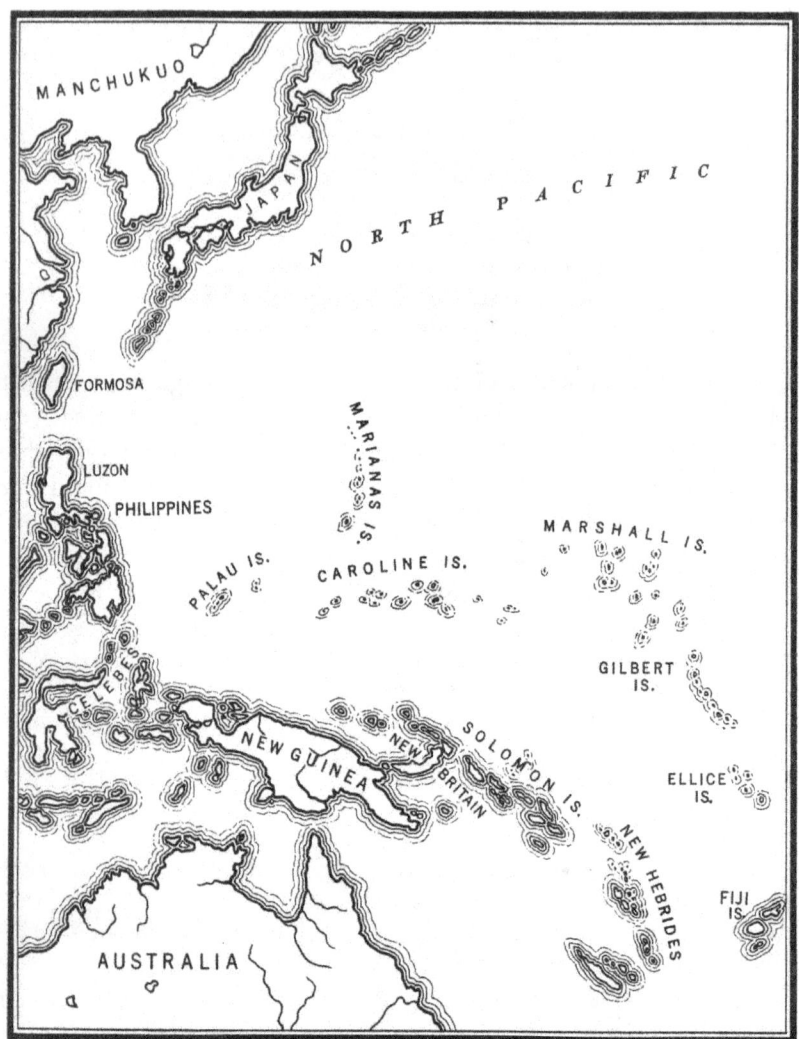

The Pacific Theater of the 37th Division

Chapter 1
A Reluctant Soldier

I WAS BORN less than one month after the Armistice that ended World War I, and I grew up on the horror stories of the War on the Western Front—especially the tales of bloody trench fighting. I had developed a strong conviction that all the maimed and dead, the Americans, the Germans, the Belgians, the French, and the British, had suffered or died to no purpose. I had come to the conclusion that Woodrow Wilson was right when he declared before the war that there were no such things as a good war or a bad peace.

In my four years at Northwestern University, from 1936 to 1940, I fell under the influence of several eloquent professors who had fought in World War I, and they furnished me with the intellectual rationale for my anti-war feelings. In my upperclassman years, as editorial chairman of the *Daily Northwestern*, class president and speaker chosen by and for that class, I exhorted my fellow students to keep out of the approaching war and to persuade their parents to vote only for the political candidates who were committed to neutrality.

In the spring of 1939, I administered the Oxford Oath ("I shall not fight in Europe") to a crowd of thousands of students at a Peace Rally assembled at the meadow of Northwestern's Deering Library. During my senior year, I organized a group of fifty college newspaper editors in an editorial alliance pledged to wield their typewriters to keep this nation out of war. My picture, along with those of like-minded editors of the campus newspapers at Harvard, Princeton, and the University of Texas, appeared in the October 7, 1939, issue of *Time Magazine*. The story included a paragraph-long account of my leadership in the cause of peace.

It may be in order here to reproduce one of several scores of editorials I wrote for the *Daily Northwestern* in November, 1939, about

Stanley A. Frankel

the war. This is a rather typical diatribe against American involvement in the then on-going battles in Europe:

> Logic and common sense are one thing; war hysteria is another. Yesterday a parade of 200,000 war veterans marched down Fifth Avenue, an avenue that many of them haven't seen since 1918 when they came back from fighting "Huns" in France, mosquitoes at Camp Dix, or submarines on the Great Lakes. The enthusiasm which these uniformed, now-graying soldiers stirred in the onlooking [sic] crowd must have been pleasing to those nations who are depending on U.S. aid in this war. It must have sounded sweet to the "boys" who now remain organized for the sole purpose of pulling political strings and obtaining pension grants. To pacifists it was an indication that we haven't learned any lessons in the past twenty years and that we are ready once again to send men overseas to defend their homes, wives and children back in Spokane.

By the time the Selective Service Act was passed in 1940, I had graduated from college and was working in New York City. Despite my strong anti-war feelings, I was not such a confirmed pacifist that I refused to register. Because my favorite Uncle Max was chairman of the local draft board in Dayton, Ohio, I figured I ought to register in my hometown. Uncle Max was my insurance against being drafted, or so I thought.

My number was among the first picked in the draft, and I received instructions to report for my physical examination. Not to worry. I phoned Uncle Max, who surely wouldn't let his beloved nephew be dragged into the Army. Wrong. He was full of congratulations, even offering that a one-year hitch would be good for me. He warned me not to be late for my physical.

I did have another ace, however. My eyesight was 20/400, far below minimum army requirements. I was, again, so sure the Army would not accept me that I told my New York roommate not to pack my things…I'd be back in a few days. At my physical exam, the Army doctor asked me to read the top line of the eye chart. When I claimed, half in jest, that I couldn't even see the chart, he laughed, patted me on the back and assured me that the Army could always find some job for a near-blind draftee. After all, it was only for twelve months. Trying a different tack, I unleashed on the doctor my outrage over the Army's carefree

Chapter 1: A Reluctant Soldier

willingness to relax its high standards for service. But he was already examining the next recruit, and I was soon off to Camp Shelby, Mississippi, to join the Ohio 37th National Guard Division training there, accompanied by my dear childhood friend, Carl Ablon, who was not drafted but volunteered because he wanted to share this experience with me.

On the troop train going to Camp Shelby, I violated one of the cardinal rules of army life and volunteered to type out rosters for the top sergeant in charge of the draftees. That "rash" act had important consequences for my whole army career.

When we arrived at Camp Shelby, I was immediately assigned to Division Headquarters to help type pay vouchers. After many tests and interviews, I was assigned to the Finance Department and then transferred to G2, Division Intelligence. Thanks to my secretarial skills, I was in such great demand that I never underwent basic military training. For me, no hikes, no calisthenics, no weapons or firing on the range. I was the prototypical paper soldier.

I quickly rose through the ranks to staff sergeant in six months, and began making plans to return to civilian life when my year of service was over. Then, in August 1941, Congress voted to extend the Selective Service Act for another six months. Though the bill barely passed in the House of Representatives, it hardly mattered. Before the extension expired came the raid on Pearl Harbor, and on December 8, 1941, my 23rd birthday, the U.S. was at war with Japan. Two months later, the 37th Ohio Division boarded the SS President Coolidge bound for the Fiji Islands.

Suva, Fiji, wasn't bad duty. For six months, my work consisted of paying the troops and typing intelligence reports. Then my G2 colonel suggested I attend the Jungle Warfare Officers' Candidate School run by Guadalcanal veterans. He promised that if I passed the ninety-day OCS, he'd pull me back to Division Headquarters where I would serve out the war as an officer in the rear echelons of the division. Sounded like another clever move, so I memorized the eye chart, passed the OCS entrance examinations, and then underwent three frightful months of infantry training. I ranked near the bottom of the 150-man class in weaponry, agility, close order drills, and foxhole digging. It didn't matter. It was common knowledge among the instructors that I'd be returning to a desk assignment, so they must have figured: "What the hell…let's graduate this infantry misfit. He'll wind up as a paper soldier anyway."

On graduation day the newly commissioned second lieutenants assembled in the mess hall to learn their next assignments. The commandant came to my name:

"Frankel, assigned to Company F, 148th Infantry Regiment."

I dashed up to the commandant and told him of his error.

"No mistake, Frankel. You were requested by Division Headquarters but there is a terrible shortage of platoon leaders, and we've decided anyone who graduates from this course will go into the infantry."

I knew damn well why there was such a shortage and wanted no part of it. "Sir, may I resign my commission?"

"You forget it or I'll have you court-martialed."

So that is how this anti-war activist, a near-blind, bumbling draftee, became a second lieutenant leading an army platoon in the war against Japan in the South Pacific. And subsequently a first lieutenant, company executive officer, captain, personnel officer and adjutant, and finally at war's end, a major…promotions awarded mainly after a battle in which the officer whom I was to replace had been killed or wounded!

The following pieces came more from my observation of what went on around me than from my direct participation. Even so, I must confess my observation point was often too damned close to the action. The events in which I was a participant, I admit, I did not volunteer for and could not avoid. But once I decided there was no way out, I did try to perform so that I could maintain my self-respect.

In the course of my 3½ years in the South Pacific, I was officially credited by the Army with having participated in five separate battles. These ranged in time from about a month on New Georgia in the Solomon Islands to a campaign in the Philippines which went on, with some regrouping intermissions, for over six months. No one soldier, including me, literally fought every day and night. All of us were shuttled in and out of direct combat, and we had plenty of time to review what had gone on yesterday and to speculate apprehensively on what was going to happen tomorrow.

During those actual firefights when real bullets and real shells and real bombs seemed to have my name on them, I had difficulty comprehending exactly what was happening to me. I could have been acting a bit part in a Hollywood war movie. Reality only intervened when the fireworks stopped and I had time to reflect. Soldiers in front of me and behind me and to the right and left had been hit and I had once again been spared. At war's end, I had my share of medals, but not one was a Purple Heart.

Chapter 1: A Reluctant Soldier

I should also add at this point an apology for some of the pejorative words which will crop up in some of my battlefield descriptions. Many of the words used to describe the Japanese may seem callous, bigoted, and disrespectful. The reader should remember that most of these pieces were written immediately or shortly after the events they describe. I would not have been there if there had not been a war on and the Japanese had not been the enemy. My feelings were bound to surface.

Remember, during combat, the only good Japanese was a dead Japanese.

I could have edited out all the mean-spirited comments, the ethnic slurs, the references to "slant eyes" and the like, but the result would have been a distortion of what was then the reality. These pieces are about a war, and wars are not easily prettified. So, I have let stand what I wrote. I trust that you, the reader, will accept what follows without being offended. I assure you that today when I find my seat companion on the Scarsdale commuter train to Grand Central is a Japanese businessman, I bear him no ill will.

The author (center) in his basic private days involving some heavy duty

Stanley A. Frankel

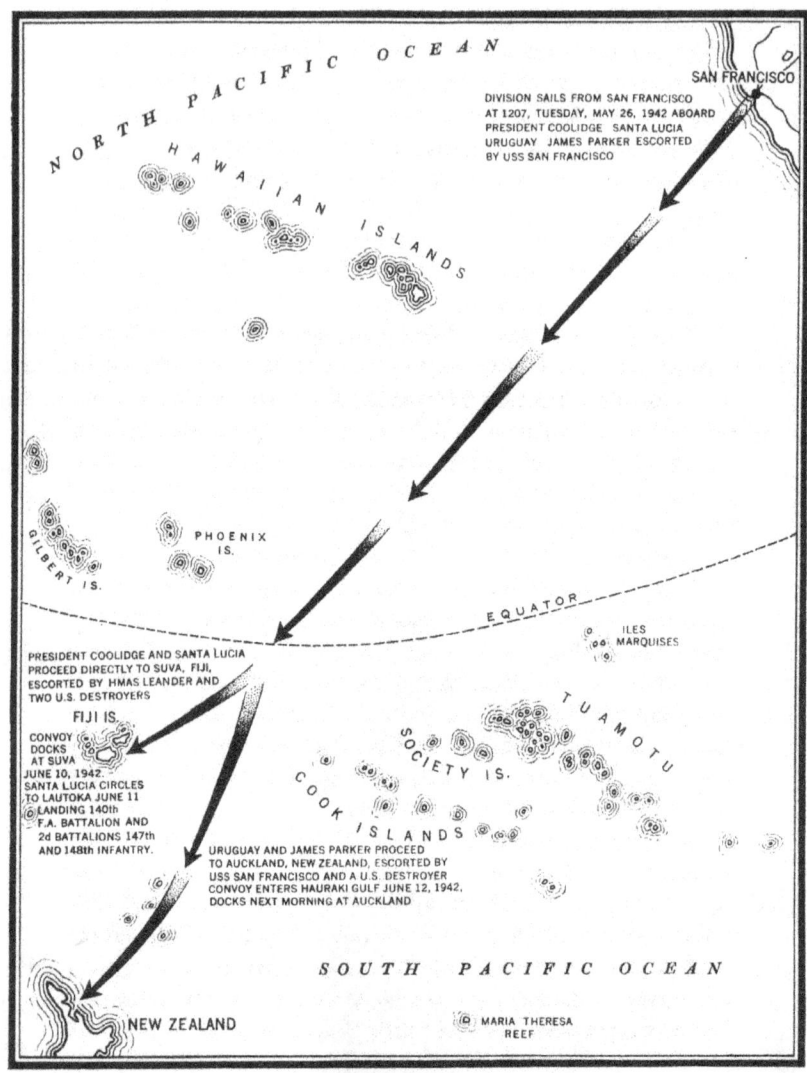

Frisco to Fiji

Chapter 2
Every Night We Die

As I've indicated, I was both a participant and an observer. I talked to soldiers returning from stomach-churning patrols and interviewed other platoon leaders for their accounts. I listened to company and battalion commanders, who added up their combat losses and who pointed out the successes and failures of their assigned missions. Some of these interviews were my military assignments, and others, my writer's curiosity, and some a little of both duty and inquisitiveness.

Once a battle was over, whether I had been a participant or observer, it was my duty to write to the next of kin of the soldiers in my company who were killed in action. It was also my job to draft recommendations for awards for the men, living and dead, whose heroic actions would be recognized by medals ranging from the bronze star all the way to the consummate soldier's recognition, the Congressional Medal of Honor.

If I had to slice out just one piece of World War II action in which I was a participant, I would choose the one in New Georgia—our first serious action. It came after a few weeks of innocuous mop-up in Guadalcanal, where the Marines and the Army had finally stopped the Japanese incursion down the Solomon Islands chain. New Georgia was one step above Guadalcanal.

The Japanese had a small airfield at the southern tip, called Munda, from which they bombed, ineffectively, nearby islands held by the U.S. forces. Our division assignment was to capture the islands and take over the airfield, from which we could then start air action over the more northward islands like Bougainville, and eventually the Philippines.

I had entitled this piece "Not so Silent Night," but when it was to be published by The Saturday Evening Post in 1943, the editor had thought that title was too soft and religious, so it was changed to "Every Night We Die." Not bad. Not inaccurate. And not subtle. Whatever the title, the writing reflects my deep involvement, my keenly felt personal experience.

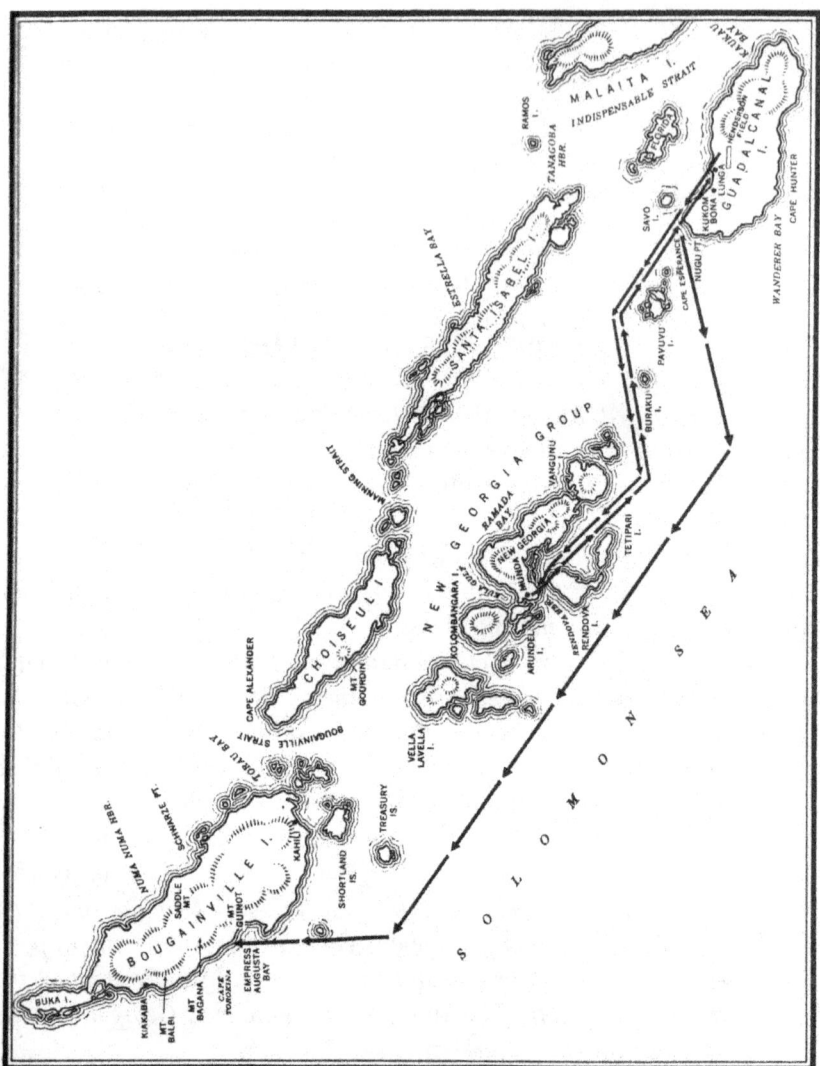

The 37th's Path through the Solomons

Chapter 2: Every Night We Die

THE PLATOON LEADER has been stewing and fretting since noon. Sgt. Kincaid's squad left two hours ago to knock out a pillbox blocking the path of the platoon, and, as yet, Kincaid hasn't reported back. The platoon's two remaining squads are pinned down by the pair of machine guns which the sergeant was supposed to destroy. The guns snap out bullets occasionally—at every gust of wind rustling the grass and at each American helmet scraping against undergrowth. Those babies have plenty of ammunition, it seems. The lieutenant's belly is raw from crawling to and from his squad leaders, trying to outguess the Japanese threat. The *crack-crack* of the weapons, however, bowls over the logic of his tactics. Nothing works.

Thank God, Kincaid reports back. He is grimy, scratched, and uncertain. Mission unsuccessful. He has outflanked the pillbox up ahead, and then he discovers two more pillboxes directly behind that one. Pillboxes leapfrogged. Of course.

The lieutenant should have known. One squad can't lick three leapfrogged emplacements.

"How many men did you lose, Kincaid?"

"None. Saw the other guns just in time to get out of the field of fire. Had to go mighty slow and careful-like coming back."

"Good boy."

The next move is obvious. The lieutenant sends his runner out to bring the two other squad leaders to him. Meanwhile, he mulls over a plan to commit the whole platoon to wind up this dirty business.

A messenger from the company commander arrives before the squad leaders can assemble.

"Four o'clock, sir, and the captain says to drop back to high ground and keep the Japanese off while the company digs in for the night."

The platoon leader nods disgustedly. Damn little done today. At this rate, we'll be here another year. His platoon has been chosen as security for the company. Its mission now is purely defensive, like the boxer's left jab, pushed out there in front of the main body to keep the opponent away and off balance. This is one plan for making the Japanese impotent during the night. The plan starts at four on the dot. It ends at dawn the next morning.

During the day, the Japanese is just an ordinary fighting man, deadly only when behind logs, in high jungle growth or inside concrete pillboxes. He exhibits an extreme reluctance to chance a bayonet, knife, or knuckle duel with our boys, and he will press the inevitable hand

25

grenade to an honorable stomach rather than be confronted by a Yank tommy gunner at shaking-hands range.

At night, he's lethal. Tricky. Deceptive. The jungle blackouts embellish his sneak tactics. The hideous fear of the dark which most of us tossed off years back has been rekindled by just one night in a Japanese-infested bush. Thus, the need for a simple, workable plan. Our scheme is: quit fighting early and get set for them.

The security platoon creeps and crawls to high ground, fans out in a semicircle facing the enemy, and just watches. Danger is over, until dark. A tired sniper might risk a try at one of the sweaty, broad backs lying below him, but he's a punk shot. If he tries to fire more than once every hour, he is quickly spotted and cut into quarters with the "fixer," the Browning automatic rifle (BAR). Two hundred yards behind the platoon, the company digs in furiously. Must be dug in, fed, and bedded down by dark. Our jungle army shuts up shop at nightfall.

The company commander surveys the diggings. Four-man foxholes are scooped out of the coral earth with hand tools, sticks, dog tags, and fingernails. The holes are shaped like a cross. Expediency, not religion. Each of the four arms of the cross is deep enough, wide enough, and long enough to permit a man to lie down below ground. No space to spare. His head is always toward the outside of the cross. In dead center, eight muddy shoes touch. When one man spots danger, he alerts the others by kicking sharply. Night signals are noiseless. Our company perimeter defense is self-sufficient; we are responsible for our own protection, but we must tie in our defenses with the companies on the left. We also must help keep our brothers out of harm's way.

At six, the security platoon is relieved in order to dig its holes on the inner perimeter and to pour down some cold meat and beans. These men will not be on the rim of the circle tonight. The extra hours of watchful waiting as a security unit earn them a concession. They'll sleep a lot better. It's a warm feeling to know that between you and the enemy are some of your own boys. At dusk, each soldier automatically crawls into his hole, places his rifle in an accessible cranny, folds his raincoat underneath him, and his shelter half over him. He lies on his back with his jungle knife, unsheathed, in his clenched fist. The Yank sleeps but his killer instinct is wide awake.

The CO and the platoon leader race around the perimeter for a final check. No sense dallying. There are Japanese out there. Our boys know the rules of this game. The good discipline we have indoctrinated into them during those days of buttoned pockets, shined shoes, and hand salutes is our insurance now. Or so we hope. And pray. The rules? Only the men on the outer rim of the perimeter will fire their pieces. They will shoot at 100% foolproof Japanese. Monkeys, coconuts, and ghosts are not to be molested. Men on the inside of the perimeter will never squeeze their triggers during the night. They defend themselves with knives, bayonets, fists, rifle butts, teeth, and religion. No one is to leave his hole. A moving object above ground is presumed to be an enemy. Knife him first and identify the corpse afterward. No talking or whispering. No smoking. No crying.

The Japanese are out to make us fire in their direction. Make the Americans start shooting. Scare them into blasting away at shadows, twigs, each other. Lie back and shoot wherever you see the flash of muzzle blast. Make them so trigger-happy they'll be murdering one another. We'll move in later and finish off the wounded. Drive them crazy by night and pick off the neurotics by day. Panic. Panic.

And on our side: keep cool. Stay put. If you see the whites of their eyes, don't shoot; jab a bayonet into those eyes. They will scream and whistle and even call your name. They will fling firecrackers at you and chant, "American, you die." But make them come in after you. When they crawl around the ledge of your foxhole, they'll silhouette themselves against the sky. Then's the time to lunge upward with that jungle knife. Keep plunging and withdrawing the hilt until the flesh is death-cold. The Japanese can't see any better than you at night. If they can't panic you, they are bewildered, beaten. When they crawl to your hole asking for it, men, give it to them. Not before.

You remember the night a slick creature crept up to the CO's hole.

In good English, he whispered, "Slip me a hand grenade, buddy."

No one had ever addressed Capt. Orville Wendt as "buddy" before, and this Clyde, Ohio, national guardsman pressed the trigger of his tommy gun and cut the Japanese in half. One minute later, a sweet-smelling companion creeps to his dead comrade. You can hear him breathe heavily in anger as he discovers that his fellow animal is dead. It's still too dark to see, but, vindictive, he feels for the hole of the man who murdered his friend. Suddenly, those fingers are on the edge of the captain's hole and the Japanese is on his knees, ready to spring. But he is uncertain. Maybe the thing in this hole isn't the murderer. Maybe the contents are garbage. He rocks over the hole, peering down into it intently. He rocks back and forth several times, indecisive. The third lean does it. Wendt presses the trigger of his tommy gun. One shot rings out and the gun jams. The Japanese jumps or falls into the hole on top of the captain, who, without any ado, beats the Japanese's head to a smashed tomato with the butt of his jammed gun, and tosses the Japanese outside. The next morning, the Japanese was examined. The one bullet had done the trick. Through the head. The bashing was so much lost motion. Besides, the captain shouldn't have fired at night. But the men disobey these rules, too, under similar circumstances. Our discipline is good, not perfect.

You'll never forget the incessant, nerve-racking *boom-boom-boom* of the artillery. All night the 105's and 155's artillery deposit their shells around our lines to fend off the concentration of enemy troops. These guns are aimed at some distance beyond our front lines to allow a margin of safety for the inevitable short shots. From bitter experience, the Japanese learn of this umbrella, and at night, at the first boom, the Japanese crawl in close to our lines, whooping and hollering as the shells pass over us and them. Americans won't come out of holes and fight at night. So they revel in their superiority. They jabber like dope fiends, and sometimes, when you imagine they are crawling closer to your hole, you grab a hand grenade and jerk out the pin. You can't throw the grenade right away. It takes five seconds for the thing to blow—plenty of time for the alert Japanese to pick it up and throw it back. So you suck in your wind, count as slowly as your nerves allow, but faster than you should, "One dead Japanese, two dead Japanese, three dead Japanese," and then you arch the little pineapple into the center of the Japanese circle. You hear screams and warnings as the little men go after the grenade to toss it back, and the split second between that grabbing for the thing and its explosion seems like forever.

You laugh when you recall your orderly, a Tennessee hillbilly who manned a mean BAR. Back in those training days at Camp Shelby, Mississippi, this lanky son of the hills was either drunk or AWOL. Just a no-good. Out here, when the chips are down, you'd rather have him along on this mission than the brightest boy in the class.

A grenade has finally blown up inside a Japanese pillbox and a piece of what was left inside cries in broken English, "Kill me! Kill me!"

Old Tennessee trains his BAR on the slit in the front of the box and calls out in that Bob Burns drawl: "Just raise yo' haid, buddy."

You remember with a twitch the night that the screamers—GI for dysentery—had caught half the company. Spoiled meat in the daily ration. The men lie in their holes, gurgling, gasping, bawling. Their stomachs knot up and they undergo backbreaking contortions trying to

Chapter 2: Every Night We Die

make a toilet out of their helmets and still remain below the level of the earth.

One kid whispers, "To hell with it," and jumps out of his hole and squats.

Zing. A bullet from our own area grazes his nose, and he leaps back to the safety of his trench like a frightened kangaroo. This shot is answered by one from the Japanese side of the perimeter. Suddenly, the tension and the misery are all relieved in an orgy of blind, wild shooting. Vampires and branches, birds and boulders all catch hell in this session of crazy, erratic firing. It lasts one hour and then stops. All night we hear a kid crying in the next hole. Next morning, we take count. This kid has been shot in the groin. No other casualties, saints be praised. A .30-caliber bullet did it. Our own. The captain blows his top, raves, threatens court-martial and violence for anyone who breaks the no-shooting rule again. He need not rave. That kid we wounded sobers up all of us.

You can't forget the mouthy guy from Auburn who was on guard in the perimeter hole and notices the bushes wiggle. He lets go with a burst from his tommy gun, and a bleeding Japanese falls out of the bushes. In his hand is clutched a pistol, probably a German Luger. Precious souvenir. Only twenty yards away. The jerks in the next hole also spy the pistol. They are fifteen yards away. The mouthy kid thinks: *those jerks will jump out at dawn and get my souvenir.* He deliberates: *won't be dawn for another hour. No officers around. I'll sneak out and get it now.* He jumps out of his hole and doesn't bother to duck as he twists the Luger out of the Japanese's hand. Suddenly, one cap pistol shot. The mouthy kid turns slowly, straightens up a bit, looks at those "jerks" fifteen yards away, tries to gesture apologetically with his hands, drops the pistol, collapses on top of the Japanese, and dies. Some sneaking son of heaven had lain in those same bushes for an hour, waiting for a sucker to come out and claim his precious souvenir.

Those ugly, repulsive land crabs—animals about the size of a volleyball—inspired more indignation than the Japanese. You recall that first night when your unit went into the front lines, fed full of weird stories about Japanese jumping into foxholes and slitting throats with devilish ease. You imagine you hear noises. You know you hear noises. A scratching sound comes nearer and nearer. It's a Japanese for sure. You kick the others and all are suddenly tense. Closer and closer comes this scratching, until you see a tiny shadow edging its way over the side. The Japanese's hand! You squeeze the handle of your knife and your body is full of moisture and pain and electricity. You wait resolutely for that next second in which you pit your knife against the Japanese bayonet or stiletto or club. Then—plop—this volleyball crab falls onto your chest and, in shock, you jump two feet off the ground. You fling the damn thing out and you lie there quivering for hours while the three other foxhole mates snicker. The next time you don't frighten so easily or imagine so blithely, but you are never sure. Is it a Japanese this time or just another miserable, harmless crab? This occurs six or seven times each night. You get to despise the crabs, and during the day, if you can drag one of them from his lair, you wreak a sadist's revenge by cutting off its claws one by one and then repeatedly plunging a knife into its hard, hideous body.

Above all, you remember a bit of heroism at night which revives your faith in man's humanity to man in this stink of war. You fall asleep in your hole this night because the day has been a rough one. You are awakened by a piercing scream and a ghastly, gurgling sound.

Suddenly one of your boys bleats pitifully, "Sgt. Paul's been slashed in the throat by a Japanese! He's bleeding to death! Someone come out and help, for God's sake!"

No one moves. Sgt. Paul will die in another three minutes. He was a good squad leader.

"For God's sake, he's dying! Can't you understand, medics? Save him!"

Even the insuperable valor of the medics cannot be expected to rise to this: risk your life ten different ways. Get out of your hole and be plugged by your own men before you move ten feet. Or if you wiggle out to Paul's perimeter position, get shot by a Japanese sniper attracted by the crying. Or, once in the hole with him, die a thousand deaths as Japanese crawl by, all night long, looking for you.

Dr. Isbin (Pappy) Giddens, from Millen, Georgia, crawls gently out of his underground aid station. "Hold your fire, boys, for just a second and let me get out theah."

He moves fast, but not like a coward. He goes to this dying boy standing up, with an unruffled dignity. He leans forward over the sergeant and sees that the jugular vein is severed. Quickly now, he sutures the vein together, but he knows that the darkness has made the job an unsure one. So he lies on top of this boy all night, his own body above ground level, holding the sutured vein in his fingers. It breaks once, but he deftly sutures it again. No Japanese comes near Pappy tonight. Don't ask us why. The next morning, Paul is given ten doses of blood plasma, is sent back to the clearing company, to the field hospital, to another island.

Chapter 2: Every Night We Die

This very hour, Sgt. Paul is leading a platoon in other jungles. Doc Giddens is still a battalion surgeon. Only two minor changes: Sgt. Paul has a deep scar in his neck, and Doc Giddens' hair is snow white.

Stanley A. Frankel

Chapter 3
The New Champions

The preceding chapter concentrated on night fighting in the jungle. This chapter covers how our soldiers conducted themselves in the unfamiliar jungle and zeroes in on a rather typical daytime foray. The American people had been brainwashed by movie versions of these battles, starring Randolph Scott or Gary Cooper. Yes, we did have plenty of heroes, but their conduct was not exactly as dramatic and fearless as that of Scott and Cooper.

This chapter was sent as a letter to "Aunt Madeleine" in June 1944. It was one of those she was unable to peddle. "The editors found it too grim," she wrote. "They are looking for heroics not 'prosaics.'"

So...let's focus on the blood routine of jungle fighting.

THE THIRD PLATOON of Company F files by in a staggered squad column formation: the first squad on the left, the second on the right, and the third, behind, for the time being. The third platoon is ahead of the whole battalion, as an advance guard, moving up the bulldozed trail to find, fix, and finish a strong Japanese machine gun position 800 yards ahead. One second lieutenant and thirty men, now.

Three days ago, one second lieutenant and forty men. Last night, at 5:00, a bulldozer operator, forging a supply trail through the jungles, turned to the left as he reached the crest of a small hill. He had two squads, a security detachment, as he made the turn. When the Japanese machine gun, located someplace on that crest, stopped firing, he had about one squad left. He didn't mind. The first burst had gotten him. The next day, when his body was recovered, his face had powder burns from the muzzle blast. The Japanese had let him come that close before pouring it on.

Two days before, the third platoon had fought a tough skirmish. They had been moving up for a week now, mucking forward during the day, and settling down in the same muck that night. The platoon leader, in between the second scout (who was thirty yards out) and the two lead squads, glanced back at his men. He wants to know whether any of them are getting ready to drop out from fatigue. More than that, he wants to see on their faces that determination and quiet courage which he himself lacks. He inwardly prays that he'll look around and be reassured that behind him are thirty Gary Coopers who will come through for him when the chips come down within the next hour.

Not a Cooper, nor a Scott, nor a John Garfield. Just frightened American kids stumbling along, unwillingly, but relentlessly. Frightened, did I say? Mostly, scared stiff. Petrified. The leader of the 1st Squad, Sgt. Wise, has lost that cocky sneer and repartee. He barely acknowledges the half-hearted expectant smile of his platoon leader. Glares back, his bloodshot eyes expressing: "On with this goddam job."

Number two man has a tommy gun at port arms. A 110-pound midget named Price, who used to tire rapidly on practice marches and maneuvers. He is bushed now, completely and irrevocably, and his uncertainty doesn't help his mental attitude. The platoon leader manages a fatherly grin toward Price. But Price wants no paternalism. He wants to lie down.

In his choir boy's voice he squeaks indignantly, "Can't... keep... this... up... much... longer." And he can't. He will pass out during the firefight that's coming up.

Belski, the fat rowdy Pole, spits on the ground and keeps his big Browning Automatic Rifle at port arms. He fondles the BAR gently. Back at Camp Shelby, and over on the Fiji Islands, Belski used to bitch about the big gun. It weighs twenty pounds to an M1's nine. The big husky boys get it, and it's the old convincer with plenty of valuable firepower. Belski knows that now. Back at garrison he used to complain about other things too: the lousy grub, the ill-fitting uniforms, the stuck-up second looies, and the goddam Army. One time on maneuvers he snuck off to the side of the road and went to sleep while his company was flanking an imaginary enemy platoon. His CO went crazy looking for him that night, called out the MP's, had all rivers checked. Belski didn't give a damn. What's a week's KP? But now he's moving up that mucky road. Tense, chilled, a soldier who knows that the little games are over, who senses deeply the whine of bullets shot from .25-caliber guns in the hands of men who want to kill him. He smiles back at the platoon leader, giving his head a little twist as if to say, "I won't let you down, sir, but I'm weak in the knees."

Chapter 3: The New Champions

Sowers, following Belski, has an '03, Springfield rifle, a lighter gun since he's an assistant BAR man and must take over if and when the Japanese start looking for our fast-spitting BAR and get Belski. He's tall, emaciated looking. Wiry underneath. Hard as his stubborn head, which caused him to be busted four months ago for insubordination. Told the new platoon leader that he knew more about troop leading in one minute than the looie did in a year. And added, "You ain't goin' to catch me doing what you say in combat. I'm on my own." Surly, then, he was a troublemaker all along. But now sheer terror is written in his eyes, and the mouth that used to curl up derisively at each command is quivering. He looks hard and long at the platoon leader as if to say, "I don't know anything except I'm scared. You've got to tell me what to do. I'll do it but I can't figure things out well here."

On the other side of the trail, moving parallel to the first squad, is the second. The squad leader had walked along the column telling the men to button their lips, watch their side of the road, and scan trees for snipers. He is vicious in his stinging orders, but the men welcome this belligerence for God knows they need it. The squad leader doesn't feel belligerent though, just sounding off to make sure his voice doesn't croak and his tongue isn't paralyzed. Gets away with it, too. The platoon leader thinks, *Good man, Thompson. Alert. Aggressive.* He marks him down for the frontal attack, if and when it comes.

An 18-year-old tommy gunner is second in the right file. He joined the platoon fresh from the States two months ago. Had been plucked out of his senior year in high school and shipped overseas after ten months of what he called "Just markin' time." He now is trying to keep his tommy gun at port arms, but the diagonal is so distorted by his twitching arms that he reminds the platoon leader, for a minute, of that drum major from Northwestern. Only the strut isn't there. Nor the smug conceit. Nor the flashy clothes. Why think about that now? A helluva thing to be reminded of by a nervous kid. The boy is self-conscious as the platoon leader stares hard at him. He knows he is shaking and he tries to answer back that it's all a lie.

"Sir, I ain't really scared." But he chokes up on the "scared," baby tears welling up in his eyes.

He looks down at the Solomon Islands muck and the platoon leader shakes his head savagely and says, "My God, is this what I've got along with me?"

The third man with an M1 has it strung over his back. He doesn't look to the right or left because he is sick from the meat and beans he crammed down in thirty seconds before moving out this morning. He gags every two or three minutes although the food is out of him by now. He is grey, grimy, in semi-shock, but he keeps on moving, putting one foot ahead of the other, like the tightrope walker at Barnum and Bailey's. This boy isn't in this world. He staggers, and stumbles. The platoon leader must snap him out of it. He's no good to himself or the men.

"Come on, Chuck, tonight we go into regimental reserve. Hot food. New shelter half. Maybe some water to shave with. No more Nips sneaking around."

An unadulterated lie, since there is no real reserve in this game. The Japanese are there in front and there in back, and they're in the middle of your perimeter, attacking you from the inside. There's no warm food or sleep for many, many nights. Maybe the platoon leader can fool Chuck, but he won't fool Belski, or Wise…or himself.

Suddenly the platoon leader sees his lead scout, ahead about 100 yards, flop to the ground and raise his rifle horizontally over his head and bring it up and down twice. Enemy in sight. The platoon leader waves the platoon off to the side of the road and crouching, moves forward hurriedly to investigate. *Christ*, he thinks, *will I ever get them started again? Wonder what this is? Awfully early for the Nips to hit us.* He gets to the first scout and tries to locate the bogeyman who the lead scout swears is in those bushes twenty-five yards to the right.

"Rot."

The platoon leader sprays those bushes with his tommy gun. The lead scout won't believe anything. The platoon leader stands up straight, a pretty target for that bogey man, and the scout is satisfied.

"The Nip must have run," he mutters.

The platoon leader grins in relief. "Even you can miss them once in a while, Jim."

Once in a while? Hell, any mother's son who can spot the little guys in their holes once in twenty tries is a wizard. The platoon leader hasn't seen a live one yet in three weeks and God knows how many have taken potshots at him. Standing upright in full view of what is up ahead, he signals the men to get up and move. Like a slow-motion movie, they get back on the road and begin shuffling forward. Wet with perspiration, hands clammy, and spines cold. The men, as they lay there off to the side, had thought of Mary, and home, and a Pabst Blue Ribbon.

Every soldier had said to himself along the trail in that jungle, "I don't want to die, and why in hell was I picked for this dirty business?"

Chapter 3: The New Champions

These were their thoughts as they waited for the *zing* and the hammer blow in the stomach, or the head. They thought this, they heard the platoon leader's voice, they crawled to their feet, and they kept on moving forward.

And kept on moving forward. There's your American boy going to war. He's a farmer, a clerk, or a shoe salesman, with a BAR in his hand. He hates the dirty job he has to do more than he hates the enemy. He doesn't hate Nips at all now because his hate has been overwhelmed by the one emotion of self-preservation. It's kill or be killed. But he keeps on moving toward the enemy. He fights the Nips, yes, and sometimes he fights them with only a tiny bit of reluctance. He has a bigger fight against his instinct, and his training, and his desires and his background. He doesn't go into battle with chin high and mouth set in a fierce come-what-may-we'll-do-it grin. His shoulders are invariably bowed with fatigue and the weight of his rations, his rifle, his shelter half, and his small arms munitions, and hand grenades—and with fear. He doesn't hear that Marine Hymn or Onward Christian Soldiers. It's just the twitter of jungle birds and squelch of GI boots. He goes to meet the enemy, terrified. But he goes.

Your boys have the real courage, the ultimate courage. The superior courage. Not the damn fool variety of a guy who doesn't know what he's up against, nor the swastika courage of a man who dies for some mustached God, nor the fanatical courage of an ignorant and illiterate peasant who knows that to die in battle is to reserve a place next to honorable ancestors in the more beautiful life hereafter. What keeps him going? I guess it's the plain, unadulterated courage which doesn't need any pep talk or false motivations. It's the courage that conquers everything before it…the enemy…the terrain…and the gut fear within himself.

A machine gun chatters and the second scout falls. Dead. Dead, forever. Something to think about. No time now to think. The lead scout, up ahead of his dead buddy, is pinned down. He indicates the general area of the gun with as little motion as is necessary. No use getting a hand knocked off by a machine gun burst. The platoon leader tells the right squad leader (remember his stinging commands) to circle to the right, and the platoon leader and the first squad move up slowly…slowly… The platoon leader gets to the dead man. No firing now. Maybe he's in the sight of a Nip Nambu gun and he shakes off the gripping feeling of eternity and crawls forward some more. A burst of machine gun fire skims over his shoulder and he finds a four-inch groove in the ground to get down lower. Then moves up again. The tommy gunner puts his finger down on the trigger and lets the Nips have a long burst. Must have figured out the right place because a few animal squeals sound in those bushes twenty yards away and there is some scuffling. The Nips run, huh? Live to fight another day? We'll see. Belski moves up now with his BAR in front of him pouring an enormous amount of lead into the area. He's standing up, not bothering to take cover, and he moves faster and faster toward the damned bush. Two M1 riflemen are with him, and the platoon leader's next to them as they smash into the enemy rathole. The machine gun is there and a little blood, and our boys still keep moving ahead. The Nips run smack into the flanking squad. They take cover, but our boys come up shooting much faster than they. Two of the Japanese drop as they try to pull the pin on a grenade, and the other two run around this way and that, screaming and shrieking until the kid who has been vomiting all day coldly leans his M1 against a tree, pulls the trigger twice, and gets two of these rabbits in the backs of their shaved heads. He leans over and vomits again.

"Four to one," exults the platoon leader, but he knows that a million Nips won't compensate for the life of this one kid. He's the kid who writes sweet letters to his mother and hot letters to his girl.

"Be home soon, Mom," he writes.

"And, baby, when I get there, what fun we'll have in the Barn," to his girl.

Sort of funny, huh? Sort of sad, too.

"Let's get going," says the platoon leader. "Let the Battalion boys check the bodies." Price has collapsed. "Put him on the road and a couple of litter bearers will pick him up shortly." Can't waste time. This isn't the end of the war. The objective is up ahead. We got real business two-hundred yards further.

"Goddam it," snarls Belski. The boys are panting and burning with thirst and nervous cramps. "Keep driving, men. On the ball." A few groans from the winded soldiers, but they reform, sort of stoically, and the third squad comes up front now with the second squad dropping back a bit. "Smash and drive. Keep moving."

Not a Randolph Scott or a Gary Cooper in the whole damn platoon. Just plain Tommy Smith from around the block. The winner and new champion.

Chapter 3: The New Champions

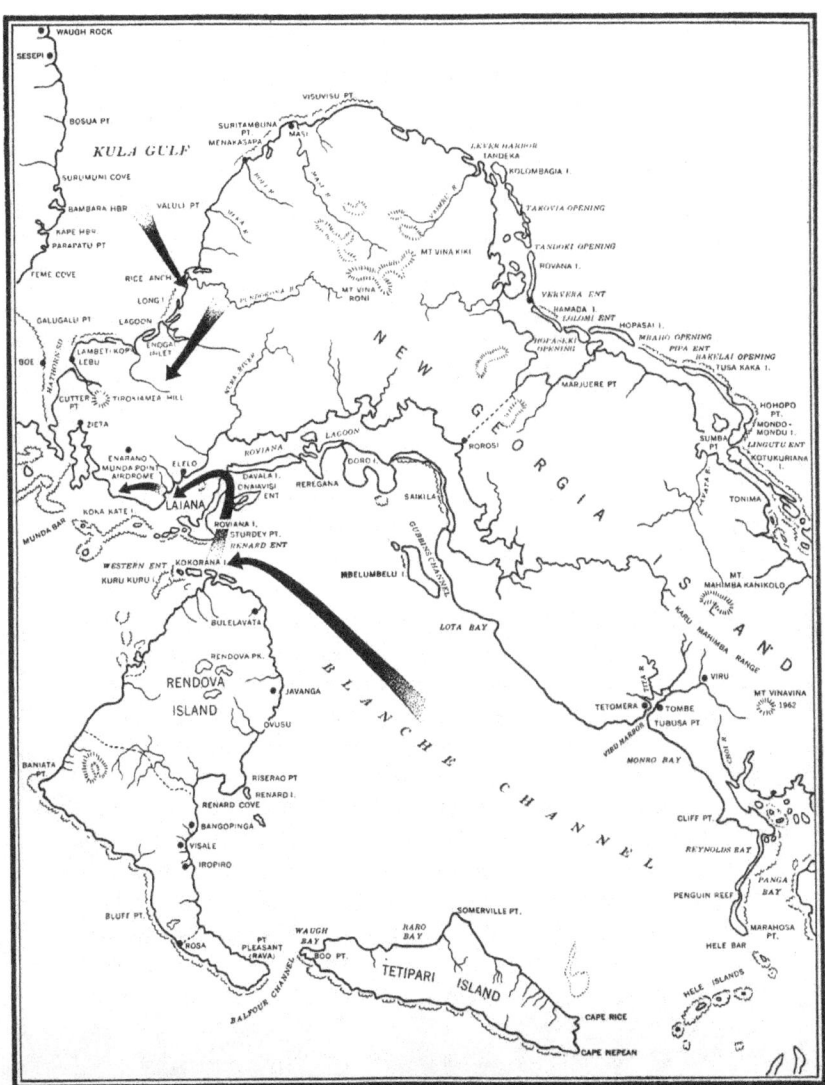

The 37th's Plan of Attack on New Georgia

39

Stanley A. Frankel

Chapter 4
The Last Ballad

M Y FIRST REGIMENTAL COMMANDER, the late Col. Stuart A. Baxter, liked to begin every speech he made about war with the line "The fog of war was pea soup thick."

If pea soup thick and fog alluded to the fact that we could not see the enemy, then the description of jungle fighting in New Georgia was appropriate. In truth, the Japanese were better jungle fighters than we were. They were better trained, had more experience, and somehow existed, even thrived, on fewer rations and ammunition, meaning they could travel faster and lighter and longer.

We would learn quickly, but the first week or two were crash course instructions at the expense of wounds and lives. On the very first day of our landing on the New Georgia beaches, Lt. Sid Goodkin took out a twenty-man patrol to feel out the Japanese. He thought he saw movement in the bushes and banana leaves to his left, knew instinctively these were American troops moving up on his flank, and called out to them that he was leading friendly troops so "Don't fire." The "friendly" troops on the flank fired anyway, one bullet catching Goodkin in the side. As he was carried to the aid station, he vehemently cursed the friendly troops for not heeding his warning. To be hit on the first day, by your own comrades, was disheartening. When the bullet was extracted from his side, he kept on cursing until the regimental surgeon showed him the bloody bullet. It was a .25-caliber slug, used only by the Japanese. The lesson was learned, the very hard way.

The patrols were sure the enemy was close only when they were fired on. The Japanese were not good marksmen so our casualties for this jungle blindness were not heavy…but we did take some. On my very first patrol of fifteen men, the company commander assigned to me a tall, husky Fijian soldier, a wise use of scarce resources since we didn't have many of these expert jungle fighters. As we moved forward, the perspiration kept soaking my glasses, and all I could see were fuzzy bushes and trees and all I could hear were birds and our GI boots crunching the ground. Suddenly, my Fijian sidekick froze...then clapped me on the shoulder and pointed in the direction of a large oak tree. I couldn't see a damn thing, but he kept clapping and pointing until, out of sheer embarrassment, I aimed my rifle at one of the branches and fired three times. I didn't hit anything, but the sniper in the tree who was about three yards from where my shots went figured he had been discovered, and he began changing positions. That movement I did see, and I aimed my M1 at the now visible target. I was shaking, a combination of excitement and panic. I fired three more times, missed completely, and the Fijian, sneering at me fired his Bren gun (comparable to our tommy gun) and brought the sniper down, dead.

We encountered no more snipers, and when we returned to our camp that night, I thought that one of these days, my life might depend upon hitting a target with my first round. So, I went to the supply sergeant, asked if I could be issued one of the few shotguns in his arsenal, and went away with a .16 gauger and a dozen shells, which I had to wipe off each morning since they, too, perspired at night.

Not too many days later, I took out another patrol, plus the same Fijian. This time I was armed with my shotgun. It won't kill anyone, but sprays pellets over a small area, usually disabling the target. That same first patrol scenario was repeated halfway through this one: the Fijian spotting a sniper, alerting me. I couldn't quite see the sniper so I aimed in the general direction the Fijian pointed to, fired away, bringing down two large leaf-covered branches...and...a shotgun-shelled Japanese sniper. As he instinctively reached for his gun to fire at us, my Fijian gunned him to death. I angrily told the Fijian we would have been better off taking him prisoner, but I don't think he heard...or wanted to.

A few days later, another platoon leader took out his patrol, which included Pfc. Rodger Young, and the account of that patrol is included here, an account related to me by the patrol leader and some of his men. Here, I was substantially an observer, but because of my writing assignments to next of kin, and for awards, I did participate in the paperwork as well as in the emotions surrounding one of our greatest WWII heroes—certainly one of the bravest in the jungles of the South Pacific.

He was the Pacific theater counterpart to Audie Murphy in Europe...except for one small detail: Murphy lived and Young died.

Chapter 4: The Last Ballad

For more than forty-five years, the approach of July 31 has caused me nightmares of jungle combat, a queasy stomach, and a drifting away of my concentration to a dank little island in the South Pacific.

For it was on that day in 1943 my little friend and comrade in-arms, Rodger Young, was "killed in action." A trite phrase...but, oh, what action!

There are literally thousands of smoky coral islands scattered like emeralds across the blue velvet of the Pacific. They've been quiet for many years.

But hidden by the unceasing jungle growth, buffeted into nothingness by the storms and the bulldozers, are the abandoned tanks, jeeps, and M1 rifles, and buckled, concrete airstrips. They've all succumbed to time and the creeping jungle. The men who ripped up those islands, hacked out the airfields, and filled the vastness of the Pacific with war, are long gone. Those who fought and died there are gone. They have been returned home—along with the living—and the South Pacific is silent once more.

But their deeds can never be silenced. They are written in records and history books and in the hearts of our people. The deed of one man who died there is perhaps more widely known than it might have been in the ordinary annals of heroism.

He may be remembered because of a song. Many Americans have heard the ballad, "Rodger Young," and have come to know the stirring words. Military bands used to march to its swinging rhythm. School children may still sing it at assembly.

Rodger Young died on a little island called New Georgia. He died in such a way that he was awarded a posthumous Congressional Medal of Honor and was chosen from among many heroes to be immortalized in a song of the infantry.

It should be enough to record that Rodger Young died a hero. But the facts show that he proved himself a hero several weeks before the fateful day that won him the nation's highest honor. Yet perhaps only a half dozen of us know the real and complete story of his quiet gallantry.

It began on a humid day on Guadalcanal in June 1943. The 148th Infantry Regiment, Young's outfit, was girding for its next objective— New Georgia Island with its insignificant but strategically vital Munda airstrip.

The company commander was busy that morning. He looked up sharply when Rodger Young, a thin, pale, and bespectacled staff sergeant, walked into his tent, saluted, and said: "Sir, I would like to request permission to be reduced to the rank of private."

It was an odd request.

"What is your reason for wanting to be busted, sergeant?" the captain asked brusquely.

"Well, sir—" The little sergeant reddened, and continued haltingly, "well, you see, my ears are going bad. I can't hear very well any more." He swallowed, and then finished in a rush. "And I don't want any of my men killed in New Georgia because of me."

The CO's eyes narrowed suspiciously. Was this a new twist in the technique of getting invalided home? "What's the matter, sergeant?" he barked. "Don't you want to fight?"

Young stiffened. "Sir," he said distinctly, "I don't want to leave the outfit. I want to go on—but as a private, so I'm only responsible for myself. I don't want to get anyone hurt because of me." His voice was thin and firm. "If I thought I'd be left behind because of this, then I'd rather drop the whole thing."

He half sold the captain, and an hour later the company doctor confirmed Young's story. The sergeant's ears were in bad shape.

"Shall we send him to the field hospital?" the doctor asked.

"No!" Rodger Young answered emphatically.

The doctor shrugged and the captain made a gruff apology. He promised to get Sgt. Young reduced to the rank of private "without prejudice," which he did the next day.

Three weeks later, the 148th Regiment (along with the rest of the 37th Division,) invaded New Georgia. The jungle was an almost impenetrable wall of vines and tangled undergrowth. The insects were unbearable, the food miserable, the water supply inadequate. At night, the 148th dug foxholes in mud and limestone. And, of course, there was always the enemy.

They crept in like animals by night, attacked, and vanished at dawn. With the invasion still only a beachhead, a good many men of the 148th were dead.

One evening the tropic sun took its sudden plummet into blackness just as fifteen soldiers staggered into the company lines. Among them they carried five bodies, wrapped in bloodstained shelter halves. The lieutenant in charge of the ragged platoon made his report to the captain.

That morning, he had taken twenty men on a reconnaissance patrol a mile in front of the lines. He had led his men along an old, seemingly deserted Japanese trail, overgrown with vines and bushes. After a futile search for signs of enemy activity, he turned back at 4:00, intending to be in his own company area before dusk.

As they trudged along the gloomy trail, the Japanese machine gun opened up suddenly and killed two men before the platoon could flatten into cover. The gun was fiendishly placed on high ground, commanding the entire area. There was no way around it; to rush it meant sudden death.

The lieutenant attempted a mass maneuver with his remaining eighteen men, and two more died.

The situation was critical. If they could not break out of the trap before nightfall, the Japanese would move in. With the machine gun cutting off the only possible avenue of escape, the enemy was in no hurry.

The men were pressed into the ground. There was only one hope. D company might hear the spasmodic fire and attack the machine gun nest from the rear. There was nothing to do but wait—and pray.

As it happens, prayers wouldn't have helped just then. Company D was too busy defending its own position to worry about a twenty-man platoon. And a little later it wouldn't have mattered.

Each of those sixteen doomed men had his own thoughts. No one knows what Pvt. Rodger Young, flattened in the scrub, was thinking. He might have been thinking of his family, or of Clyde, the little Ohio town where he grew up. He might not have contemplated that he was only 25 years old, which is pretty young to die. Or he might merely have been thinking that the omnipresent Japanese machine gun was a nuisance—and dangerous to boot.

What went on behind Young's spectacles and between his rather deaf ears no one knows. What is known is that he began to inch forward, cradling his rifle in his arms, past the lieutenant and toward the machine gun nest.

The lieutenant saw him slither by, and tried to grab his leg. But Young was in a hurry and evaded his superior's grasp. Furthermore, the Japanese saw the rustle of grass and loosed a burst that singed the lieutenant's hand and tore his collar.

"Come back here!" the lieutenant screamed at Young. "It's suicide. Come back—that's an order!"

Young hesitated a moment, then twisted his head around and grinned at the lieutenant. "I'm sorry, sir," he said, "but you know I don't hear very well."

He turned then, and continued to snake his way toward the Japanese emplacement. They saw him coming, of course. A stuttering burst cracked into Young's left arm and splintered the stock of his rifle.

Young let the useless weapon drop. Still, he pressed forward. His buddies fired blindly at the emplacement, trying to divert the spitting stream of death. It didn't work.

Another burst of fire sewed a scarlet seam down Young's left leg, from thigh to ankle. But he kept going, and finally reached a shallow hole about five yards from the machine gun. It was deep enough to afford him rather tenuous safety as the Japanese apparently couldn't depress the muzzles of their guns far enough to get a clean shot at him.

"For God's sake, Young," the lieutenant shouted, "stay where you are! We'll get you out somehow."

Maybe that time Pvt. Young really didn't hear. He might have been dying at that moment. In any case, he wasn't in the mood for playing possum.

Painfully, he reached into his belt with his good right hand for a grenade. He pulled the pin with his teeth. Then, rearing up and back—up out of his position of relative safety—he lobbed the grenade toward the machine gun.

The gun answered with a blast that caught him full in the face. Rodger Young died as the grenade left his hand. Still, well thrown, it lit in the center of the machine gun crew—and killed every one of the five Japanese manning the weapon.

Within seconds, the fifteen survivors were on their interrupted way back to Company D. Silently they carried their five dead. Rodger Young didn't need to worry any more about being responsible for the lives of his buddies.

Two weeks later, the sweltering little island was in American hands. The troops stopped hunting and killing and began collecting themselves—and some well-earned medals. Company D's captain composed a lengthy recommendation that Pvt. Rodger Young be awarded the Medal of Honor. One sentence read:

"Disregarding the orders of his platoon leader to come back, Rodger Young moved forward into the face of enemy fire." The regimental commander changed that to "Not hearing the orders." No one in *his* regiment disobeyed orders, he remarked acidly.

The company commander also wrote letters to the next of kin of those in his unit who had been killed, including, of course, to the parents of Rodger Young. A month later, Rodger's mother replied, thanking the captain for his note, commenting that Rodger's bravery made the loss of an only son a bit easier to bear, and asked for a small favor: "Rodger was proud of being a staff sergeant. Since his body won't be returned to us until after the War, we would like to put up a little monument in our Clyde, Ohio, cemetery, and would it be permissible to write 'Staff Sgt. Rodger Young' on the tombstone?"

The captain quickly put in a request to Division Headquarters, asking that Pvt. Rodger Young be promoted posthumously back to staff sergeant. He recounted the original reason for the demotion, the posthumous Medal of Honor, and attached the request from Rodger's mother. The Army remained inflexible. The request was denied; a number of army regulations were cited justifying the turndown, and when the captain protested in a personal visit to Division Headquarters, he was told firmly that there was no appeal.

Furious…whipped…he had to write to Mrs. Young that "for the time being, your wonderful son would have to remain a private."

But this bitter rejection turned into an amazing benefit. Most of the branches of the military service had songs, but the infantry did not. The War Department assigned songwriter, Frank Loesser, of "Guys and Dolls" fame, to write a piece for the infantry, suggesting he read the Medals of Honor citations for inspiration, and instructing him that he must only focus on infantry privates. Not sergeants, not officers, but privates!

Frank Loesser sought out the single most dramatic act committed by a doughboy-private. The Rodger Young story was shown him and he had to read it only once.

A military citation is a strange place in which to find inspiration for a ballad. But from just such a dispassionate source sprang the moving ballad of Rodger Young. On the face of it, it is a song commemorating the gallantry of one soldier. But when you hear it sung, you know that there is more than one Rodger Young, just as there are many islands in the South Pacific that knew a Rodger Young:

"Oh, they've got no time for glory in the infantry,
They've got no use for praises loudly sung,
But in every soldier's heart in all the infantry
Shines the name, shines the name of Rodger Young

Shines the name—Rodger Young—
Fought and died for the men he marched among,
To the everlasting glory of the infantry—
Lives the story of Private Rodger Young."

It is plain that the name, and the story, of Rodger Young will live. And as the years go by, the fact that he was small and rather spindly, the fact that he needed powerful glasses, the fact that he asked to be demoted because he was going deaf and did not want this disability to jeopardize the lies of his comrades—these will fade from memory. His name and his story will long outlive such details, and will become inseparably merged.

For today, long after the last trace of war has vanished from those quiet Pacific islands, Rodger Young has taken his place among the legendary heroes of American history. And there he is quite at home.

Chapter 5
To The Rescue

One of my most gut-wrenching experiences in New Georgia (with a happy ending) came out of my diary and my subconscious as I reconstructed the action forty-five years later:

FOUR PERSPIRING LITTER-BEARERS lugged Sgt. Copenhaver into the Regimental Aid Station, using a ragged shelter half for a litter. He was terribly shot up, and the blood puddled in the shelter half and seeped through the tiny holes, dotting the ground with red. With a disciplined gentleness, the litter-bearers laid him across two medical supply chests that had been rigged up into a surgeon's workbench. Capt. William Welker, the regimental dentist from a little Ohio town, expertly needled him a shot of morphine. Then he probed for the vein in the soldier's arm through which the plasma would flow. The wounded man was conscious but in shock. He had been slugged in the back, the right leg, the right arm, and the buttocks. His right ear, cheek, and neck had been nicked. There, the blood dripped, didn't spurt.

Copenhaver was the driver of our company's three-quarter-ton truck which I had sent to the 37th Division water point, about 2,000 yards back from our CP, to fill twenty-five five-gallon cans with desperately needed water. That was two hours ago. The five guards I had sent along were not back. Only Copie.

I tried to talk to the sergeant as he lay there moaning and staring, with the doc holding the plasma bottle high above him. "What happened, Cope? Where are the others?"

He twisted his head back and forth a few times and bit his lip, then that blank look. I rounded up the litter-bearers and plied them with frantic questions. Copenhaver's truck had been ambushed about 1,000 yards away by an infiltrating Nip patrol. The men in the truck, said the litter-bearers, must have died fast. Cope had literally been shot out of the driver's seat, nosediving into the mud. The others just slumped over where they were sitting. The truck had begun to burn, and the Nips, whooping and hollering, had run to the cremating bodies just to make sure. Probably had poked bayonets into the doubtfuls. Copenhaver feigned death, but had to bite a hunk of mud to keep from groaning. The Japanese kicked him a couple of times, and, satisfied, took off. Cope was able to pull himself to the side of the road and drag himself 100 yards in the direction of the command post. The litter-bearers stumbled across him and pieced together his story of the grisly picture nearby.

The six guards? Dead. I winced, coughed, choked. A sledgehammer cracked in the pit of my stomach and kept pounding. This was the minute I had anticipated and dreaded for the past five months. I knew on maneuvers in Fiji and on routine patrols on Guadalcanal that this would come. Yes, the time was here, five days after we had landed for our first real brush with the enemy. I had sent six men to their death. Six soldiers died carrying out a mission that I ordered. Good platoon leaders get killed and bad ones get their men killed. Yes, bad platoon leaders get their men killed.

Whenever I had blundered over the crest of the hill back in the practice maneuvers on the Russell Islands, my commanding officer had laughed at me for taking it so much to heart. "This is just a dry run," he used to say.

I knew better. I realized that such blunders cost lives in battle. The payoff is a .30-caliber bullet in the chest, not a little chagrin and wisecrack. Here it was. Nobody's fault. Could have happened to any officer. Occurs every day, every hour. One of those things. No consolation. Why hadn't I gone on that detail myself? The supply road was reported safe. How could I know? The Nips always sneaked through our flanks and harassed the supply line. Why wasn't I there when the Nips cut loose? Better to have a few slugs in the gut than to live with this the rest of my life.

Today it was my men who were the statistically reckoned 3% who were destined to die. Three percent, hell. Those kids weren't percentages. They were Pvt. Roy Wenland, the lovable, garrulous tub-of-lard who could talk his way out of a KP assignment with the *savoir faire* of a One-Eyed Connolly. And Pvt. Max Aaronson, a tiny Jewish lad with an MA from Columbia University who taught the fourth grade back home and never once complained about his low rank in the army. They were Beans Simpson, the pale, shy farm boy who wrote his mom every day in a childlike scrawl, so damn tough for us censors to read. The same letter all

Chapter 5: To The Rescue

the time: "I'm good. Hope you are the same. I love you, Mom." Awfully corny, huh? Pvt. Scottino, with a BS in Economics and a reflective gripe about everything from the taste of coffee to the smell around the latrine. Pvt. Elmer Underwood, an Arkansas hillbilly, who was good for more laughs accidentally than the USO comedians elicited on purpose. Pvt. Oscar Anderson, an officer's orderly, who worked in the Bureau of Internal Revenue in Washington before the draft. He could talk rings around the colonel, but was from hunger [sic] when it came to shooting a gun. These six equal three percent? These guys weren't statistics, they were men with whom I had eaten and slept, whom I had drilled, led, insulted, commanded for the last five months.

I stumbled around the CP like a man spaced out on drugs. My stomach was upset and my head spun a little. The war had finally struck Stan Frankel.

Just then, Capt. Edward Nicely, the Regimented S2, yelled for me, and I sleepwalked over to him. A phone call had come through from the wire chief. He and his wire-laying crew were caught in the same ambush area, cut off by the same Nip patrol which had hit my boys. There were just five of them in the crew, and by the grace of God their wire line was still intact, and they were able to beg for help.

Nicely excitedly told me to round up the rest of my platoon and "get 'em."

My platoon couldn't be located. The men had been assigned to special missions, patrols, outposts. I was able to rout out six men: two officer's orderlies, an old supply sergeant, a grounded truck driver, and two of my tommy gunners. They responded slowly to my frantic orders to get their tails moving. I yelled insults, screamed at them to follow me. Didn't even explain the mission. If I had, they would have taken more time.

We started down the Jeep trail, and I half ran for several hundred yards. I stopped, looked back, and found I had outdistanced my cautious crew. I commented loudly on their yellow streak, but thereafter kept my own pace down, retaining a tiny bit of discretion. At 1,000 yards we ran smack into the fatal clearing. In the middle was this burnt-out truck. Contrary to the litter-bearers' report, the men had tried to jump out of the truck, and they sprawled all around the side, stiffened by death into crazy poses. I thought I recognized "Fat" Wenland with his heavy jowls in the mud and his limp right hand on the running board. He was facing the safe side of the road, shot in the back.

Stanley A. Frankel

We stopped cold. The odor of burning rubber and men cleared my emotionally unbalanced head like a whiff of ammonia. I was completely sober. The Nips might still be over in those bushes to the right, waiting for the suckers who were sure to come after their dead. Fishing with human bait. Maybe, please God, they had departed on another murder mission an hour ago. I didn't know. And if they were still there with their slant eyes patiently searching the avenues of approach, I couldn't tackle them with my small crew. Even if we made a run for it, some of us would get shot. I mulled over sending back for more men. Would have been logical. Safe. Those dead men up ahead said no. We had five men to save, and a delay might add five to five. To hell with stalling. And, then, it would be embarrassing as hell if the Nips had actually taken off. So I conceived a plan, a Boy Scout plan. Simple, but the best I could do. I told my tommy gunners to move around to the right, circle to the Nips' flank, or what would be their flank if they were there. They were to take good covered positions and start firing like hell in the direction of our little yellow brothers. Fifteen seconds after they opened up, the rest of us would dash across the clearing, hoping that the Nips were bluffed into thinking that a heavy force had discovered them and was closing in. Or, hoping that they would be diverted for five seconds…five seconds for us to get across the clearing. Then the tommy gunners would run like the devil back to the CP for help while we five found the wire chief, dug in with him, and helped fight off the Nips until that help came.

The tommy gunners stalked off, and our hearts pounded in unison for the next five minutes. Suddenly, that *crack-crack-crack* which we sometimes confused with the Nambu light machine gun. This was ours. Fifteen seconds, and then we took off. I dashed like a goosed rabbit across the clearing and even the old supply sergeant found a bit of youth and passed me up as we made the other side. We were safe. We had outfoxed the enemy. Maybe. Maybe we had shadow boxed with ghosts…but at least we had made it. We kept on at top speed until we ran into our "embattled" wire crew. Only they weren't so helpless now. A whole 200-man company, the advance guard of a battalion moving up from the beach, was all around them. Warrant Officer Elmer Hyter was bitterly cursing a bulldozer operator who had torn down his wire lines. Belligerent now. A couple of his crew were sitting by the side of the road munching C rations. I was damn glad to see them, especially surrounded by so many of our "allies." I hadn't relished a firefight with my pick-up crew, most of whom were better at draw poker than fire and movement. And frankly, I was no one-man army myself.

Chapter 5: To The Rescue

While I readjusted my pack, I blinked about four times, rubbed my eyes, and then jumped for joy. It was "Fat" Wenland, probing a few cans of C rations in order to mix hash with meat and beans. This boy was alive. He was fingering that chow with the old-time vitality. I ran over and kissed him on both cheeks. (Later Col. Stuart A. Baxter, of Toledo, Ohio, my regimental commander and a two-war hero, told me with a wink that this was conduct "unbecoming an officer.") Wenland ducked back defensively and said with his eyes: "The looie has gone bats."

"Where are the rest, Wendy? Are they still alive?"

"Naturally," intoned Wenland in between half-mouthfuls of hash and beans. "The boys are over behind that big tree playing rummy with Scottino's marked deck."

I dashed over to the tree just in time to find them arguing heatedly over the appearance of a fifth ace.

"Yippee!" I hollered exultantly.

They all stared at me incredulously.

"Combat fatigue" whispered Scottino out of the side of his mouth.

Here was their story: Copenhaver had left them at the water point to take a much-needed bath while he delivered a few stragglers and the water to the CP. He was to pick them up in an hour. When he didn't show, they started walking, worried about catching hell for being late. They joined the wire chief, Hyter, and decided to stay with him until the road ahead was clear. They weren't taking any chances.

We trekked back to the CP. On the way we loaded the four dead boys on our own shelter halves and lugged them back to the CP for Chaplain Wareing to bury. This didn't dim my elation. These four were the three percent. Not my boys. These four men were what the next day's communiqué would mention in "Our losses were light." My guys were all right. We strolled back to the CP and reported in to Capt. William Leathers, of Hornbeak, Tennessee, a fighting man who used to drop baskets in for LSU...and who swore by Huey Long. The whole company crowded around. Not only had Lt. Frankel "rescued" the wiremen from the enemy—he had also brought six other of our boys back from the dead.

Leathers commended me as much as he ever did: "Christ, are you back alive?"

I'm not modest, but I had to disclaim any resemblance between me and Colin Kelly. The men would have none of it. Several weeks later when the fighting had subsided the adjutant, Capt. Cotterell, tried to build my 2,000 yard round trip walk into an award recommendation.

After unearthing the real facts, he trimmed the recommendation down a bit and it later became a combat infantryman's badge.

Copenhaver was evacuated that evening. Before he left I told him the men were all OK, and he smiled.

"Hell, lieutenant, I could have told you that. Did you ever see Fat Wenland get into anything he couldn't talk his way out of?"

Of course not. He was one of my boys. Not any damn statistic.

Chapter 6
Taking it From the Air

Here may I back up for a few minutes, back from New Georgia to those rather passive pre-New Georgia actions which began on Guadalcanal. Our first taste of combat, a kind of appetizer, had come after we had left the Fijis for mop-up action in Guadalcanal. The Marines had done their messy job of winning that island, and our mission was to find and kill the few remaining Japanese stragglers. The days on patrol were easy, but the bombing every moon-filled night was scary. Here's an account of a bombing raid:

We were awakened about 3:00 this morning by the chilling sound of the siren. We don't sit up in our cots because our instincts are well disciplined. Flat in bed, our bodies are below the level of the ground. We live in holes. Sitting up, our heads would extend above the earth's surface and we would become vulnerable to the Japanese favorite aerial treat: the one hundred pound daisy cutter which slices off a blade of grass at its root for fifty yards around the impact area.

The searchlights easily spot the enemy bombers, just two of them, and they are caught in half a dozen crisscrosses. The offbeat motors are unmistakable. Mitsubishis. *Thrub Dub. Thrub Dub. Thrub Dub.* Washing Machine Charlies. The *ack-ack* starts and the *pom-pom-pom* of the 40 millimeters plays a duet with the *crash-bang* of the 90s. The .50-caliber machine guns pop away harmlessly. Their tracers fall far short of the target, up there about 3,000 feet. The Japanese at first circle around warily until the lights catch them. Then they cease their waltzing. We can just make out the opening of the bomb bay doors, and the *ack-ack* music is suddenly drowned out by the tremendous *whooshwhoosh* as bombs fall diagonally over our heads and beyond. Inaccurate as usual. Harmless as usual. The earth shakes again and again as the bombs, in a string, hit the ground a second apart. The planes start out to sea; the *ack-ack* dies down reluctantly; the searchlights stretch and stretch until they can no longer follow the movement of the fleeing enemy. We go back to sleep.

This was either our fiftieth raid or our hundredth. Too many to cause much excitement anyway. Many of us sleep right through the bombings. Since we live underground and since we have enough overhead protection (two layers of sandbags) to protect us against falling *ack-ack*, we feel relatively secure. Some of us jump a bit when the first anti-aircraft gun explodes nearby, and when we hear the *whoosh-whoosh* of the falling bombs, we have that fleeting suspicion that this one has our serial number inscribed thereon. We are pleasantly amazed, however, at our casual acceptance of this suspicion. When the bombers turn tail and scoot for Rabaul, we go to sleep easily. That is, unless the Spam we had for supper conspires with our digestive juices to keep us up.

One year ago, it was a different story. We'll never forget our first raid, a bombing attack which completely shattered our nervous system even if it shattered nothing else except a Lever Brothers Coconut grove. That raid left us queasy for days. We had gone into Guadalcanal just after the Army and Marines had settled the Henderson Field issue. It was ours to keep for good. The lone Japanese recourse was the airplane and until we arrived at the Canal Washing Machine Charley had come over every night.

We unload off the transports into the Higgins boats and hit the Beach at Kokombona, Guadalcanal, about 4:00 in the afternoon. De-limbed coconut trees, rusty ammunition, and a few stinking, half-buried bodies are all around. We are sobered up by the smell. The CO points out a cleared area 200 yards inland and instructs us to dig in for the night. We need no pep talk. This is the real thing. This is not a simulated situation or a maneuver problem. Tojo is coming over about midnight. So we think, and we dig well and deep. Of course, we do not get a raid.

Two whole weeks pass without even a "condition red." We move from one campsite to another and our bomb shelters become progressively shallower. We get very smug.

"Here at the right time," the men laughingly say.

A gentleman's war. Japanese air force knocked out.

We finally get our permanent camp and we erect pyramidal tents and pitch our cots. We scratch out a little earth as a pretense for a bomb shelter and devote our main energy toward the construction of a luxurious, open-air log-seat movie theater about a quarter of a mile from our "home."

Chapter 6: Taking it From the Air

We go to the picture show this Wednesday night. It begins at 7:30 and the moon is starting to come out full again. Good. We don't need to lug our flashlights and precious batteries with us. There are 500 soldiers altogether at the show, half our own battalion and the other half a black port outfit.

The feature attraction is "Take a Letter" starring Rosalind Russell and Fred MacMurray, and we are beginning to enjoy MacMurray's predicament as the male secretary when it happens.

It happens fast but I can recall every detail. First, the siren wails.

Someone moans: "Condition Red."

Ros Russell hurriedly departs from the screen. One second later we hear a strange sound in the sky, this *thrub-dub*, like a broken-down washing machine. It's unlike the sweet purr of our own B17s and P38s. The searchlights shoot upward and the beam catches this silver bird overhead, directly over our heads it seems. This is it.

Someone mutters: "Breaks our virginity."

Then we watch bomb bay doors open and we notice little "sticks of wood" fall from those bays.

I estimate that two seconds elapse from the siren to the falling bombs. We had been frozen for those two seconds, but the sight of the "sticks of wood" coming earthward thaw us out in a helluva hurry. We tear away from the theater area, scanning for ruts, open latrines, mudholes.

I am an officer and I should do something to ease the pandemonium. I remember that responsibility only after I dive into the relative security of a blessed watery ditch. Then, ashamed, I pray that my commanding officer, a rough and vitriolic captain, hasn't witnessed my cowardice. Not for long, however, as I look underneath me and there he is, quivering and perspiring.

The bombs hit the ground just as we dive into the ditch, perhaps three seconds after we started to run. They land a few miles away. We should have known that when a bomber is directly overhead, its cargo of death can't hit you. We are embarrassed, and when the all clear sounds, we walk back to our seats with our eyes on the ground. The show resumes. Fifteen minutes later we get another condition red. This time, a little less confusion, a little longer to find holes, a tiny bit more courage, and the skipper and I regain face a bit as we shepherd the men into their ruts before walking hurriedly into our own.

For two weeks, the bombers come over four or five times a night. We get no sleep. We are either in our foxholes while the bombs are dropping, en route to our beds when condition green is given, or lying in bed in-between worrying about how fast we can put on our shoes and scoot to our holes when the inevitable wail of the siren warns us of approaching danger. We desperately fear the Japanese sneaker who might come in over the mountains undetected and blast us to damnation while we are dreaming of Hedy Lamarr and Mary Smith. Each time we hear those bombs *whoosh-whooshing* we feel certain that this is the one which has our number etched in its nose.

At the end of these two weeks, the officers and men of our regiment are semi-neurotics. No bombs have actually crashed within one mile of our bivouac area, but that doesn't help our mental state. We keep assuming that the law of averages will start working and we will get our share soon. We officers are disgusted with our reactions and we try to discover some solution to the jitters which have infected us and thus doubly infected our men. At last, each of us evokes a palliative. I convince myself that while I'm in my hole, nothing can hurt me except a direct hit. If the hit is direct, then I'll never know what happened anyway, so why worry? Furthermore, I'm not fighting the war by myself. This is a big island and there are thousands of men and foxholes on this strip of land. Chances are about one in a million in my favor. The jitters are licked, cold!

Outside of the intense fright which I had experienced back then, I sensed deeply two other emotions. The first was that of intense hatred, hatred of the plane above me which forced me to cringe and grovel like less-than-a-man. Hatred of the little sonofbitches flying that plane who had the audacity to zoom over an American Army camp with murder and contempt in their hearts. Hatred of the whole Japanese military machine from Tojo down to his buck privates. I wanted to see our night fighter knock down those intruders more than I have wanted anything else in the world. One thrilling night we witnessed the black P38 swoop down upon a bomber which had been enmeshed in the lights. The lights flickered off and we could see the tracers from our avenging angel rake the silver bird from propeller to tail. We leaned out of our holes, ecstatically entranced. The minute the bomber exploded in air and the big ball of fire came floating gently down into the sea, every man on the island was out of his hole, screaming violently in fiendish glee. The cheering reverberated for miles and each of us would have kissed that kid pilot, our guardian, our protector. I felt the same sensation a couple of centuries ago when I saw Ernie Lombardi win a ball game in the last of the ninth with a heroic home run.

Chapter 6: Taking it From the Air

The second emotion was effusive admiration for the little people of London, Malta, and Chungking. Here we were, hardened soldiers, quivering cravenly under the pinpricks of a couple of cheap Japanese bombers. There they were, women and children, stubbornly bearing the brunt of the most powerful, until then, air attacks in history. We were ready to hit anyone who sloughed off the bombing of London as a commonplace and unheroic event. Ridiculing those English civilians for their courage was the same as spitting in our eyes.

Right now, during these sporadic, ineffective raids, many of our "damn fools" get out in the open and watch the *ack-ack* and the night fighters and the Washing Machine Charlies as if they were looking up at a baseball game. The rest of us average guys remain in our bunks, secure and self-confident, mentally thumbing our nose at the *thrub-dubs*. When the inevitable *whooshwhoosh* comes, we still get that fleeting suspicion: "Here's the one with my name on it." But, what the hell's the difference?

Between engagements on the beautiful Guadalcanal beaches

Stanley A. Frankel

Chapter 7
Japanese Soldiers, Docile and Cooperative

THE MOVEMENT of our troops across the small island of New Georgia, toward the Munda air strip along the Pacific Ocean, was slow and bloody. But it was inexorable since we outnumbered the Japanese at least 5 to 1 and had superior weapons and supplies, while they had begun to run out of food and ammunition. After three weeks, our sweep led to the capture of the airfield and the driving of the surviving Japanese into the sea. On that last day, about 100 of the enemy waded into the ocean, and as the water got deeper, they threw away their guns and began swimming toward God knows where. There were no islands closer than twenty miles away and no friendly ships to pick them up. They just swam to a watery suicide.

I quickly learned first hand what the expression "shooting fish in a barrel" meant; our infantrymen, joined soon by artillery troops, quartermaster corpsmen, and anyone else close by, stood on the beach shooting at the flailing enemy. As the few faster-swimming Japanese reached spots beyond rifle range, the Americans hauled up 60mm mortars with instant-contact shells and lobbed them in the area of the lead swimmers.

Division Headquarters heard about this and jeeped down several Nisei[*] translators and a loudspeaker system. The Niseis positioned themselves in the middle of the firing soldiers and called out in Japanese: "Come back. Drop your weapons. Raise your hands. You will be well treated."

The objects may never have heard the message. None made any move shoreward. More likely, even those who heard had been trained not

---

[*] These Nisei were bilingual second generation Japanese born and educated in the U.S., serving in the U.S. Army.

to be taken prisoner, and death by rifle or drowning was the only option they knew…or could choose.

By sundown, there were no more specks in the ocean, the shooting gallery was closed, and the fighting was officially over. This hari-kari episode is a particularly vivid example of the Japanese unwillingness to be taken prisoner. They had not been trained, as were we, to surrender in case the situation became hopeless. We were lectured that the Geneva Convention required prisoners to give their captors only name, rank, and serial number; nothing about units, size of force, tactical intentions. I do not know whether captured Americans observed these conventions; I'm sure their Japanese captors did not honor them.

The Japanese soldiers were brainwashed never to surrender: go down fighting; take an American soldier or two with you; and as a last resort, hari-kari…suicide with a bullet or a grenade or a sword. They were sternly reminded that to be taken prisoner was the worst possible disgrace, that if captured they were no longer to be considered Japanese. They were told that when war ended, that if they returned home, they would be tried and probably executed. The bad news for us was that we could not take many live prisoners; the good news was that if and when we lucked onto a petrified Japanese or an unconscious one, upon interrogation he would tell us everything he knew, even if that meant betraying his compatriots.

Anyway…the soldiers at Guadalcanal had managed to capture a few, and the following story of those few reflects their mental and physical state—some months after they had surrendered.

This word picture of the Japanese prisoner arose out of a rear echelon, morale-building, baseball game.

Our regimental team took on the Island MPs. Our boys had just returned to a rest area after the New Georgia battle, and commanding officers were trying desperately to eradicate the picture of dead Japanese and watery foxholes from the minds of our soldiers. Our regiment had several good softball teams. In spite of rules against officers playing on enlisted men teams, I accompanied the team I organized to their games, both as manager and as third baseman.

The island MPs had a topnotch team, and we traveled a few miles by 6x6 truck to their well-kept diamond. Both sides warmed up a bit and then settled down for the ball game. Pvt. Gaines was up first and drew a base on balls. I was up second. I took a good hefty swing at the first ball, missed, and it was just then that I noticed *them*.

Chapter 7: Japanese Soldiers, Docile and Cooperative

They were sitting behind some chicken wire fences about twenty yards away, parallel to our third base line. Our team bench was between the line and the chicken wire. About twenty laughing, cheering, jabbering Japanese were watching our game! The same species of men that we had killed in New Georgia and whose maggoty bodies we had smelled soon after they died. The identical features, builds, skin. Just better fed and a lot more jovial.

Needless to say, trying to watch the ball with one eye and these Japanese with the other didn't work. I fanned.

One of my men forgot himself for a second and called out disgustedly: "Keep your goddam eye on the ball…sir."

I walked back to the bench and pointed out my discovery to the rest of the men. They ceased to be interested in the game. We gawked at these Japanese while they smiled back at us and cheered us, unconcernedly. Later we learned most of them had been prisoners for about a year, having been captured at the Canal. As far as we knew, these were the same as the enemy who had shot at us from trees in the Munda fight, who had ambushed our medical trucks on the Jeep Road, and who had held grenades to honorable stomachs when the jig was up.

We lost the game 1-0. The third baseman bobbled two easy grounders, and the Japanese screamed lustily. The louder they screamed, the more he bobbled. It was an odd experience. One week ago we had shot our last little yellow brother, and this afternoon he was cheering us on to victory at his favorite game, baseball. The whole business aroused my curiosity because I assumed that all prisoners had been evacuated long ago. Also, twenty Japanese is a helluva lot. Our division took about six prisoners during the whole campaign and all except one were half dead with malaria and multiple bullet wounds when we grabbed them.

The next day, I returned to the stockade and was able to round up an old OCS chum, Lt. Jones, who was now in charge of these prisoners. He told many tales and permitted me to interview them and to nose around. He explained that no publicity had been given out on this little Japanese detachment, but he saw no objection to writing home some of the pertinent information.

Jones explained that these men were ideal prisoners. After the first day of internment, most of them understood that they were not going to be tortured or starved. In fact, the C rations and the cigarettes lavished on them were luxury items. They bowed to their guards and they bowed to the medics who ministered to them. They bowed to little religious books and dolls they had strung around their neck, and they seemed overwhelmed at the kindness of their captors. In return, they cooperated in every way: never attempted to escape; forever volunteered for work details; and gave every bit of assistance they could to aid in our war effort.

The most literate of all, a Japanese Zero pilot with a rank equivalent to that of our flying sergeants, spent two full days interpreting a Japanese document which had been picked up in a cave. He didn't go to sleep for forty-eight hours until the questioning was complete. The information which was gleaned from this document was of value in our intelligence estimates. He must have realized he was being a traitor, but that didn't seem to faze him. This Zero pilot had been the newest acquisition, having been picked up several months back when the Allies went into the Treasury Islands just to the south of Bougainville.

The prisoners cultivated gardens, built mess halls, latrines, and kitchens, erected their own barracks, built tables and lounge chairs, bought cards, ate heartily, laughed, played baseball and basketball. One thing they did not do was write home, even though the Red Cross offered them facilities to do so. In fact, they blanched and shook their heads violently whenever they were offered a chance to inform their folks that they were safe. They were prisoners of the enemy, therefore dishonored, and Nippon had severed all relations with them.

Lt. Jones explained there was one infallible method of convincing a reluctant Japanese prisoner to tell all he knows: to tell him, via interpreters, that they will send his name and picture back to Japan. At that, the Japanese falls to his knees, begs forgiveness, and proceeds to spill enough beans to send a few squads of his brothers-in-arms to honorable ancestors. We might note that few Japanese ever attempt to hold anything back. Being taken prisoner is not in their handbooks. No Japanese is ever taken alive. Thus, they are not drilled in the "name-rank-serial-no-nothing-more" routine. They usually reveal everything easily without any persuasion and seem unhappy when their lack of information does not permit them to answer a specific question. For this reason, our men have been repeatedly warned that though all dead Japanese are good Japanese, a live Japanese prisoner was worth ten times his weight in American lives. The bravado with which some American soldiers cold-bloodedly shoot quivering, defenseless, even wounded Japanese, in this context, could be equated to treason. The moral aspect of the situation was secondary; the significant point as that taking Japanese prisoners could save American lives. If one of our soldiers killed a Japanese he could take prisoner, he was in effect putting a pistol to the

Chapter 7: Japanese Soldiers, Docile and Cooperative

head of three of his buddies. If the prisoner had enough information to lead us to one camouflaged pillbox, he had saved arms, legs, lives.

I talked to the Zero pilot, Lt. Tasha. He was small, handsome, personable. In broken English he answered every question politely and enthusiastically. He worked in some kind of ordnance plant as a civilian in a small town near Tokyo. He had a Japanese high school education and had studied four years of English. He was inducted into the Imperial Army Air Force, and applied for pilot training. He was accepted and in the summer of 1939, earned his wings, and won his right to pilot a Zero. First sent to the Chinese theater, he accompanied bombers on routine missions. He never saw a Chinese plane in the air and had no encounters with the Flying Tigers. In fact, he never actually pulled the trigger of his machine gun except to clear the gun or to do some practice shooting. He spoke earnestly, apologetically. Maybe he didn't strafe women and children. He was convincing. He returned to Japan in the summer of 1941, got a new plane, and did some flying with a fighter squadron. When the Japanese launched their Pacific drive, he was sent first to the Philippines after the fighting had subsided. Next he went to Truk, then Rabaul, and when the Americans seized New Georgia, to the Kahili airdrome on Bougainville. Tasha had never engaged an American plane in combat and was forced to land on the Treasury Islands when his motor went haywire. The island was then in Japanese hands but fell a short time later and he was taken prisoner, without putting up any fight. He had great admiration for the American P38 but retained a reverence for his own Zero. He didn't expect to return to Japan ever, had completely excluded his family and his wife from his thinking, and appeared pleasantly resigned to going to the States and working there even as a prisoner for the rest of his life. Tasha shrugged his shoulders when you asked him who was going to win the war. In the back of his oriental mind were still those many years of indoctrination: "Japan has never lost a war. Singapore, Hong Kong, Corregidor. Greater East Asia ruled by the God Emperor."

Kori, the five-foot ten husky, was next. He had three middle teeth knocked out, a crew haircut, and the build of a wrestler. Kori didn't laugh, but not because he wasn't happy. He was the workhorse of the whole crew. In Japan he had been a farmer, had been drafted into the army at the age of 21. The interpreter estimated that Kori was one of the reluctant dragons for he frowned when asked about army training. The work was hard and the food wasn't as good as on the farm. He too traveled to China and engaged in several skirmishes. He claimed that the

Japanese never lost a battle in China; often raided a town and then withdrew without seeing a Chinese soldier. Kori was sent to the Philippines when Japan invaded those islands, and was in a reserve Division which didn't have much action. He saw a lot of American prisoners. His superiors told him that the war was about over, that the Japanese had taken Hawaii, and that San Francisco would be invaded next month. He was dispatched from the Philippines to reinforce the Hawaiian garrison (which explained to him the long ocean voyage). He didn't know it wasn't Hawaii until he was taken prisoner by a U.S. Marine officer who told him that this was Guadalcanal, an island of which he had never heard. Since that time, he was up on his geography. He blamed everything on his officers, claimed that the Japanese soldier was not brutal or warlike, that he would like to surrender, that he loved life. But the officers were vicious, using whips, gun butts, and fists to beat iron discipline into the ranks. The men were violently afraid of the officers, even more than they ere afraid of the Americans. When the Japanese lost their officers, they tried to surrender but the Americans wouldn't let them. He swore that Marines shot ten men in his company who advanced forward with a white flag almost to the Marine lines before being fired on. He disclaimed any knowledge of Japanese faking surrender and tossing hand grenades. With that, I took a couple of salt tablets.

The other prisoners substantiated a great deal of what he said, especially the terrifying fear of the officers. Our own experience with their officers was substantiating. An officer would not be taken prisoner. He would kill his men and then himself. If taken prisoner after being shelled into unconsciousness he refused to talk and was insolent and swaggering. The Japanese enlisted soldiers regarded him with awe and fear and, while he was in the room, refused to answer even the simplest questions. When he left, they gave him fierce glances. It was some time before the men could be convinced that he could do them no more physical harm. Then they talked but they kept looking quickly toward the doorway through which he left.

Chapter 7: Japanese Soldiers, Docile and Cooperative

A Japanese Zero plane downed in New Georgia

Stanley A. Frankel

Chapter 8
Once Upon a Christmas

After our regiment was moved back from New Georgia to Guadalcanal to prepare for the next assignment, I had some free time in between administrative work catch-up. Many of those spare hours were spent writing stories about our recent engagement. They were mostly factual—as factual as a writer with poetic license can get. Occasionally, however, I would bang out a roman a clef—a piece of fiction based squarely on fact. One of these stories was about an imaginative Christmas spent in the jungles. As it turned out, this piece was pretty close to the reality of our next Christmas, and it could stand symbolically for all infantry soldiers in all wars fighting in jungle-like conditions on December 25.

A MONTH AGO, our boys departed from New Georgia. The fighting was all over. We had helped take the airport. Before New Georgia it was the Russell Islands. Before that, Guadalcanal. We've spent one Christmas on the Fiji Islands, so a tropical Yuletide will be no novelty.

We are now behind the front lines, somewhere. We are reorganizing, getting set, shadow boxing with machine guns, doing roadwork in jungle terrain, sharpening our punches on a 200-yard rifle range. We can't promise that we'll be at it again December 25. We don't know. I'm writing this as if a foxhole Christmas is the prospect.

That's why there'll be little outward sign of holiday. We are ground-bound infantry. We will be too close to Japanese lines to make even crude church services feasible. There will be Christmas, however, in hearts and minds. When you lie in a hole built up by bamboo logs and coral and the shrapnel whines overhead, you become converted by that rude shelter, far more than you have ever been by lofty cathedrals. The man ahead of you in an elongated squad column is knocked down by machine gun fire. You want to believe that there is someone up there who is looking out for you in future ambushes.

The cross-like four-man holes are dug, the outposts are drawn in, the fields of fire are cut, and the cans of vegetable hash are opened. Tastes pretty good, even if this same C ration has been your breakfast, lunch, and supper the past month. The boys will talk about Christmas turkeys, and the sadists will recall sumptuous Christmas meals, a hundred years ago, it seems, lingering gently on each bit of drumstick and each cranberry.

A few boys will break the rules. They'll whisper about old times, about other Christmas Eves. The captain will tell them harshly to "keep your mouths shut," but the CO doesn't know who's talking and he isn't allowed out of his hole to find out.

What do the boys think underneath this small talk? You guess it. I believe in each soldier out here there's that insatiable yearning to be home, safe and sound, warm and dry. They don't whimper, but they'll complain about being away so long.

Will it be a white Christmas by 1944? Or 1945? Or 1946? Do they pray more fervently on Christmas Eve? I guess they do. It's a holy night and that might give a little extra consideration tonight. As the enemy planes with their weird off-beat motors fly overhead and the searchlight beams finger them in the sky and the *ack-ack* tracers (like Christmas lights) shoot up and the bomb bays open and spit downward, then a fellow gets a little desperate and prays tonight or any night.

Sleep. Day soon breaks. The platoon sergeant reads the gist of the notes he took at the CO's meeting. Objective is three miles away on the knoll of that hill. The men look on bitterly.

One frowns: "Merry Christmas!"

We start our approach march with one squad preceding the main advance guard. The enemy dual-purpose gun shells us a little, more a nuisance than with any lethal value. We plod on slowly, each file looking toward its side of the jungle for machine guns and snipers. Suddenly a shot like a cap pistol cracks out. We hit the ground, wiggle toward the brush along the road, and look and wait. We can't do anything else. We think of home and Christmas again. The point squad has spotted the sniper. Lots of firing. They made sure he's dead even if he won't fall out of the tree. But there's a call for litter bearers and Cpl. Thompson is lugged off with a shoulder wound. He's glad he's out of it for a time.

We find a pillbox, surround it, throw in everything. Finally roll a couple of hand grenades into the slits. We collect our wounded. One boy is dead. We hang around and help dig the grave while the chaplain

Chapter 8: Once Upon a Christmas

scurries forward and finishes it off. It's Tom, the blond kid, whose letters I've censored for the past six months.

No one cries, because this is old stuff. Carl, the dead boy's foxhole partner, looks up at the lieutenant and kind of apologizes for sentiment: "No more Christmases for Tom, eh sir?"

Stanley A. Frankel

Chapter 9
No Rest Areas

Shortly after the battle for New Georgia ended, I was selected by the regimental commander to attend a Chemical Warfare School in Brisbane, Australia. I guess this was an honor to be the lone "selectee" in the entire division, but I was more grateful for the social respite—the clean clothes, the warm showers, the flush toilets, the beer and "steak and eggs," and the women. In two weeks I learned all I needed to know about chemical warfare. In fact, as the war turned out, there was no need to know anything since neither side resorted to poison gas. Though both we and the Japanese brought gas masks into combat, both sides quickly discarded the masks; the trail we followed to pursue the enemy on New Georgia was littered with these masks.

After the two-week respite in Australia, I bummed a plane ride back to Guadalcanal where our regiment had been shifted from New Georgia. The ride is described in this excerpt of a letter to Irene:

JUST AS WE WERE in the middle of a stretch of the Pacific, the emergency alarm rang on our plane, the plane's motors went off, the plane started getting lower and lower until it was at water's level, and the crew ran to the escape hatch and tossed out a rubber lifeboat. All of us passengers thought the end had come. I saw Eddie Rickenbacker (in mind only) and said to him, "Here I come, boy." Then, the plane veered upward, and the crew explained: another plane had fallen into the ocean; its crew was swimming out there and we happened along just in time to see them, throw them a lifeboat, and we circled over the spot several times. I saw four men get safely to the boat. We wired their position to the nearest port, and by now, I imagine they are safe and sound."

On reaching our Guadalcanal Headquarters, I was confronted with some good and bad news. The bad news was that we were immediately to train for our next Solomon Islands objective, Bougainville, the northernmost terrain. The good news was that I had been named regimental adjutant, promoted to captain, a promotion which moved forward more briskly than most paper because it was designated (not exactly accurately) as a battlefield promotion.

More significantly, my new title as regimental adjutant meant I would remain behind, in command of the rear echelon, while the rest of the regiment attacked Bougainville. In conjunction with other 37th Division troops and Marines, they were assigned to push into the island only deep enough for an airfield to be carved out along the beach. From this airfield, we could then bomb Rabaul, New Guinea, and neutralize the Japanese attacks on our sea lanes, clearing the way for air and sea dominance en route to the Philippines.

My role was now administrative: keep the records, guide the supply line, receive the wounded, bury the dead, train the replacements who would be stopping off for two weeks before being sent on to the front lines in Bougainville.

Col. Lawrence K. White, regimental commander, asked me on the eve of the departure of our troops whether I felt sorry to be left behind. We were close friends, about six years apart in age.

At 30, he was the youngest regimental commander in the Pacific, and had served under MacArthur in the Philippines before being called back to the States just prior to Pearl Harbor.

I told him, in all honesty, I felt guilty about not going along since I had been a part of the fighting until then. On the other hand, I didn't object to not being shot at or bombed.

Laughing, he prophesied (as it turned out, rather accurately) that I might see more action in the rear than the troops who were going to attack the Japanese.

Since the mission was not to drive the Japanese off the islands, but to set up a perimeter around an airfield and force the Japanese to come at us, this was a much more innocuous assignment than in New Georgia, where we had had to slash our way through the jungles and the camouflaged enemy.

Our troops landed on Bougainville, along with a Marine division, in early November. They pushed inland against little opposition, for the Japanese were obviously expecting them to fight across the island, through the jungle terrain, as in New Georgia, where they could nibble them to death. Instead, after our troops had covered an area about five miles square, they stopped, built perimeter defenses, carving out fields of fire to their front and side, zeroing in their artillery to the edges of the

Chapter 9: No Rest Areas

cleared out jungles, and settled down to wait for the Japs to come to them. Meanwhile the Navy Seabees built the airfield which it was the Division's mission to protect. From it our air forces could attack, Japanese island strongholds further north as well as a portion of the Philippines. So there was a pleasant lull from November to March until the Japanese finally understood the limited nature of the mission and were forced to come after our troops.

Meanwhile, back at Guadalcanal, I ran into more action, and more danger, than our whole regiment had experienced in Bougainville, the ammunition dump on Guadalcanal, which supplies the entire South Pacific with bullets, shells, hand grenades, mortars, rockets, flame throwers, and gasoline suddenly caught fire and blew up.

The most graphic account of that event might be my deposition at a hearing conducted to determine the reason for death by drowning of one of my recruits during that explosion. I was in charge of the investigation and my own statement was supported by many other witnesses. Mine was quite detailed, and it was accepted by the investigative body as accurate. Here it is, verbatim from the transcript report:

> About noon, November 26, 1943, while eating lunch, we heard small arms fire coming from an area approximately 500 yards from our headquarters, in the direction of the Ammunition Dump. We thought nothing of it, but slowly, both the noise and the intensity of the explosions became louder. At 1:30 p.m., the noise was deafening, and one of the men came to me with a large piece of shrapnel which had fallen next to his tent. I made up my mind to evacuate the area immediately, not because of the immediate danger but because I didn't know for sure how large this dump was, and I wanted to take no chances.
>
> I fell in both the Replacement Company and the Real [sic] Echelon, with their helmets and instructed Lt. Holloman, leader of the 1st Platoon, to march the men to the beach, walk along the beach (away from the ammunition dump), past the dangerous area, to the clearing about 500 yards away from there. I instructed him to disperse the men to await there for future orders. At that time, this new area was well out of the danger zone.
>
> I made a personal inspection of each tent after Holloman and the other officers had marched the troops off. I found a few sleepy stragglers and promptly booted them toward the beach,

on the double. By this time (2:00 p.m.) the explosions were growing louder and two shells burst almost simultaneously in our area, throwing a few of us to the ground. The men needed no more urging, and when I was satisfied that the are[a] was completely evacuated, I followed up the beach.

On the road between our CP area and the CB's, there were hundreds of men from all units, natives, CB's, marines, Australian, and Army and Navy troops. Warrant officer Bailey took the initiative of directing traffic at this point and he made it possible for the ambulance to get through on a mission somewhat farther down the road. I walked hastily toward the area where our troops were assembled, and the shells then began to sprinkle all around us. There was nearpanic [sic] among the men along the beach and complete panic among the natives. I ran up to Lt. Holloman and we agreed that we had better get the men on further down. About two hundred yards away, the Tenaru River runs into the Pacific Ocean. The mouth is about 50 yards wide, and about twentyfive [sic] yards of this mouth was about six to seven feet deep.

When our men arrived at this crossing, there was much confusion and panic. About six hundred in all were crowded around the narrowest part of the opening and many more were jumping in, attempting to swim across. Men were jumping on top of each other, and I noticed that some who could not swim were hanging onto others who barely could. By this time, the whole area was being showered by light shrapnel, and the explosions were growing in intensity. The only escape was across the river, since there were enough bomb shelters for only about 200 men around that area, and it was hazardous to return to any area toward the Ammunition Dump.

Seeing a few men floundering in the water, I asked for volunteers to go in and help them, and about five of our own 148th men went in with me and dragged a few to safety. I thought we had all of them. I then asked everyone to keep still, and I said that the men who could swim well could swim across the mouth. I had noticed a native edging his way toward the ocean, then back in, in a semicircle around the mouth and I thought there must be a shallow sand bar so, I yelled that everyone who could not swim to form a chain and I would lead them across this bar. The chain was immediately formed, and we struck out in the

Chapter 9: No Rest Areas

general path that I remembered this native had taken. Feeling our way, we were able to cross this mouth with water going up no higher than our necks. After I reached the other bank, I told the men to keep moving up the beach past Hq., Forward area which was well beyond the danger point. I stayed on this bank until the chain of about 400 men had crossed safely. I noticed about fifty men lying flat on the other side of the bank, and several officers were attempting to get all of them on the chain, but they refused. I assumed they were afraid of the water. One Captain called to me that he would find cover for the men who were afraid to cross. When Lt. Drewry and I were both satisfied that the men were either safe on our side of the stream or safe in bomb shelters, we proceeded up the beach.

I walked to Headquarters, Forward area, around which many of our men were grouped, and I talked to the Adjutant General, Lt. Col. Pruden, telling him our story and making arrangements to feed, house, and clothe our men who straggled in. I saw Lt. Harral, placed him in charge of our bunch, told him to gather the men together around Hq. Forward area and wait for me. I bummed a jeep ride to Div. Hq., reported the situation to Col. Moore, and then made arrangements with the 145th, the 129th, the Artillery, and Hq. Co., to put our men for the night…allotting 75 men to each unit. I went back, told Harral to march the men to Division Hq., and made a jeep drive along the beach for a period of three miles rounding up stragglers.

At 5:00, we had accounted for about half of our men, and after feeding, Lt. Holloman reported to me at Division and told me that he had remained on the other side of the bank, helping the men find cover. He told me that the bulk of our men were returning to our area. It seemed that the explosions had subsided quite a bit, but I was reluctant to let the men stay in that area overnight. I went with him to our area, found a lot of duds and shrapnel all over the tents and open ground, with little material damage done. A few hunks of shrapnel had torn tents, ripped cots, and shaken up the inside of the tents, but our installations and records were safe. Nightfall was approaching and since the road we would have to travel to Division was studded with unexploded shells, I ordered all men to get on trucks immediately. Three men and three officers volunteered to remain behind as guards. We set their cots in the large dugout near the mess hall, and then we returned to the Division CP., carefully signalling [sic] when a shell was in our path. The men were well treated by the various units, and they were fed, bedded down, and blanketed for the night.

That night, the night of November 26, rumors started to come to us at Division that a man had drowned. There was nothing certain but I tried to contact the 20th Station Hospital and found the line had been burnt. It was also impossible to travel there via jeep since the MP's had guards to keep people away. I learned later that the 20th Station had evacuated its patients to the hills, anyway. I couldn't sleep that night because of worrying about that report.

The first thing the next morning (November 27), Mr. Bailey, 1st Lt. Phillips (MC), and several enlisted men went directly to the spot where the men had tried to swim. The natives were diving in that mouth to bring up helmets, clothes, shoes, etc. One of them said he thought there was a body there, and Dr. Phillips went in, dragged out the body, and brought it to the 20th Station Hospital. The man was identified as Pvt. Donald B. Keeney, ASN 33505724, a replacement who had been tentatively assigned to Co. M., 148th Inf. Regt.

The men began drifting back, and Sunday morning, November 28, all were accounted for. I was so dubious that only one man could have been killed in this situation, that I called roll myself, and personally accounted for every man.

I named Lt. Holloman to act as investigating officer in the death, and he proceeded immediately to question members of our unit and examine the available evidence. Lt. Drewry was appointed to inventory the effects of the deceased.

The platoon leaders from the 148th Inf. Regt., Holloman, Drewry, Pesosky, Farrington, Harroll, and McHugh, did an exceedingly able job in cutting down the panic and assisting in rescuing scores of men. They obeyed all of my orders, and they not only exercised control over their own men but over most of the 350 non-148th men as well. I wish to commend all of them, plus an unnamed number of non-coms for their work, in helping the 600 men emerge from this situation with so few casualties.

There were many men hit by shrapnel and by diving to the ground, but no injuries were of a serious nature, and the 20th Station Hospital and our own doctors rendered effective

Chapter 9: No Rest Areas

treatment in all cases. Keeney was buried immediately by the 20th Station Hospital because of the decomposing state of the body.

The tragedy of one boy being killed has upset me tremendously even though I am certain within my own mind that all of us did everything possible to extricate every man from the trap. I think it miraculous that our loss was so small. In addition to the one man from the 148th, one CB was killed by shrapnel on that beach and an MP had an arm blown off in the same area. Light cuts and abrasions were general throughout the men in that vicinity.

A road sign in Bougainville reflecting
the indignation at the Japanese
for bayoneting our wounded.

79

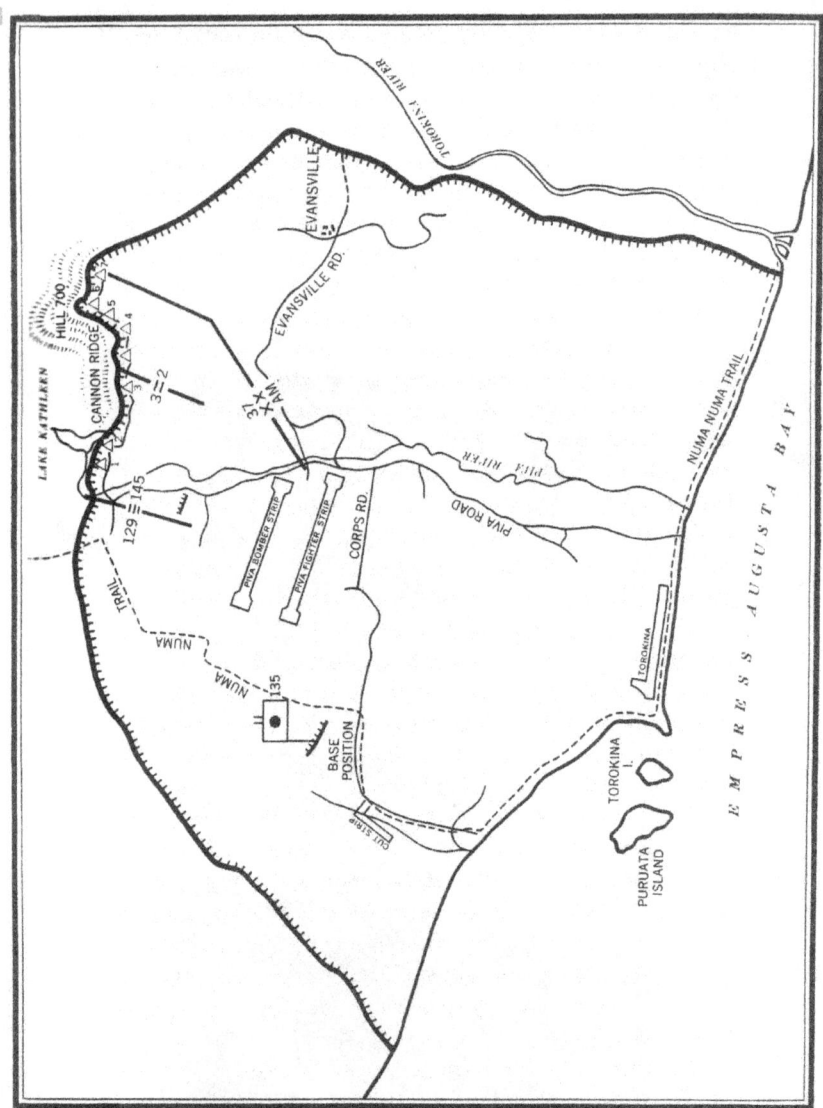

The Empress Augusta Bay Perimeter

Chapter 10
Bougainville Mopup, and Ready for Philippines

DURING THE FOUR MONTHS our troops fought in Bougainville and I commanded the rear echelon in Guadalcanal, I did make a few in-and-out boat trips to the front and happened to be there when the famed Battle for Hill 700 was fought, for which my old Company F won a Presidential Citation.

The author after landing on the beaches at Bougainville

Stanley A. Frankel

The Japanese had finally concluded that we were not going to pursue them, so in March, they had to come after us. They massed about 20,000 troops along a 7-mile-long perimeter, manned by both the 37th and the Americal Divisions. Concentrated attacks on thinly held lines always result in some initial success. One thousand Japanese suicide soldiers attacking on a 100-yard front will break through, suffering immense casualties, but counterattacks by our reserve troops who had pinpointed the breakthrough can—and did—drive the Japanese back.

We were able to anticipate the heavy attack on Hill 700, one of the higher and best-fortified sectors, by a fluke. One of our patrols, the day before the attack, probing to determine the location of Japanese troop concentrations in front of our lines, had a firefight with a Japanese patrol. They wiped out the Japanese with the exception of a lone soldier who had been knocked unconscious by a concussion grenade. He was brought back to Regimental Headquarters, came to, recognized he had been captured, and was thus, in his own mind, forever disgraced.

For the future, he was no longer Japanese, and when our Nisei interrogators asked him questions, he had been so badly prepared to be taken prisoner that he seemed anxious and relieved to supply the answers. One question had to do with the assembly area of the troops who would soon be assaulting our position, and he was shown a map of the entire perimeter. He circled an area behind Hill 700, and told the Niseis that was where approximately 5,000 Japanese were ready for an attack at dawn the next day. I sat in on this stunning interrogation, and was involved in discussing with the regimental staff whether the information might have been correct…or was it just a ruse? All those in on the questioning agreed that this Japanese had actually written off his country and his fellow soldiers and was well prepared mentally to help us. On the basis of the prisoner's information, our artillery and air force shelled and bombed the one mile square area behind Hill 700. Later we discovered the chopped up bodies of almost 4,000 enemy, many having been hit so often by so many bombs and shells that the area had the appearance of a massive hamburger facility. The smells emanating from that one-mile-square area of mincemeat, rotting under the tropical sun, were unbearable. When we sent in troops with bulldozers to bury the chopped-up dead, we finally made use of our gas masks…the first and last time they were worn.

Before that sickening experience, the Japanese, still alive, about 1,000, attacked our lines, broke through for a few hours, were then counterattacked, and driven back. The few hundred survivors ran to the

Chapter 10: Bougainville Mopup, and Ready for Philippines

other side of the island and never fought again. They spent the remaining months and years there, raising enough food to survive and building deep enough bomb shelters to withstand the periodic reminders that the U.S. forces knew they were still there.

One heroic incident which took place in our Co. F. and Co. G. counterattack that drove the enemy off Hill 700 was contained in a letter I wrote Irene:

> Two of our companies did a magnificent job in one of the counterattacks. One of the two was my old Co., Co. F. Its CO, one of my best friends, a brilliant kid named Goodkin…24 years old… performed a feat of heroism which has been unrivalled thus far. He led the attack personally in the teeth of the Japanese fire. Jumping into an unoccupied pillbox, he noticed some shells burning, ready to go off. He got ready to go out when suddenly one of his men collapsed into the hole, wounded badly. Goodkin knew he couldn't carry the man out of the hole without both being killed, so he grabbed the smoking shells and tossed them toward the Japanese with his bare hands, burning himself badly. He still wouldn't quit and led the final assault which annihilated the enemy. Only then were his raw, bleeding hands noticed, and he was immediately hospitalized. Check Finn was here at that time and Check and I went to visit Sid. I didn't know the whole story then but guessed he had done well despite his modest declaration. A few weeks later I learned that he had been recommended for the Congressional Medal of Honor, the country's highest award, by the commanding general of the Division himself. He is now with his company. His promotion to captaincy came through two days after mine so I outranked him…as far as rank goes…but I don't come up to his knees in terms of courage and leadership. If he gets the Medal and goes home to receive it, I'll tell him to call on you.

The "Check Finn" mentioned above was a Navy lieutenant, the best and oldest friend I have in the world, since the time our mothers wheeled our baby buggies together along Lexington Avenue, Dayton, Ohio. The odds that this friend might possibly be on board a supply ship, in Bougainville, the same day and night I was there, are about 2,000,000 to one, but we did manage to learn of the other's presence, made contact, met, and among other things (including hugs and kisses) visited the front lines and the field hospital together. Fifty years later we gathered with other childhood friends in Dayton, Ohio, and Check confirmed the general veracity of my war stories when our other friends began looking incredulously at their coward-pacifist pal spinning such bloody tales.

A one in ten million chance: I ran into my oldest, dearest childhood friend from Dayton, Ohio. Chester Finn (right) commanded a naval vessel that pulled into Bougainville with ammunition for the 148th Infantry Regiment.

However, most of my participation and observation about Bougainville was from afar—Guadalcanal—and may be best summed up by the following paragraph, written by the historian Stanley Frankel, not the combat infantryman:

The invasion of Bougainville in early November of 1943 was an outstanding success, and comparatively easy due to some very astute advance planning which put our forces ashore where the Japanese least expected them, and thus caught them off balance. The Marines did a fine job and pushed rapidly into the island to the southeast, while the 37th Division moved to the northeast. At that time the Marines caught the rough stuff, while we had comparatively little opposition in our sector. The Marines were pulled out, commencing Christmas Day 1943, and were replaced by the Americal Division. There was a comparative lull until March, when the big Battle of Bougainville took place. This was the attempt to the Japanese 17th Area Army, of which the infamous 6th

Chapter 10: Bougainville Mopup, and Ready for Philippines

Division of Rape of Nanking fame was a part, to drive us from our beachhead at Empress Augusta Bay. The 6th Division was by all odds the finest fighting unit encountered by the 37th Division in all of its campaigns. Here the great strength of the enemy was hurled against the 37th's more than seven miles of thinly held front. Never before had more frightful or bloody fighting taken place in the Pacific. For more than a month the Japanese smashed themselves, time after time, against our front, ultimately losing more than 10,000 killed and an un-estimated number of wounded. They ran up against a division of veterans that time...a division that proved as aggressive and powerful in the defense as it had in the New Georgia offensive campaign. We were beginning to feel the weight of more and better equipment by now. We had more air support, more and better tank support, more artillery, and, above all, men that knew the business of jungle fighting from A to Z. We refined our policy of letting machines fight for us to the maximum. For instance, we shot up more than 450,000 rounds of artillery. The dividends that helped pay is exemplified in the fact that we killed Japanese at a ratio of 33 to every American soldier lost.

The men had reason to be proud of themselves. They had fought two heavy campaigns within a period of eight months and they had won the praise of the highest Pacific commanders. Well under four hundred 37th Division men had been killed in both operations, and nearly 12 thousand Japanese soldiers had been destroyed. The Japanese hold on the Solomons was finally and completely broken. This, tied in with the success in New Guinea, spelled the end of Nipponese ambition in the South and Southwest Pacific. It was a job superbly done.

I moved my rear echelon command to Bougainville in March, and most of the next few months were literally fun and games: championship softball tournaments, inter-regimental boxing matches, volleyball, ping-pong, chess. I managed to come up with some tennis rackets, balls, and net, and we paid the natives to stamp out a playable court. Col. White and I were the two most enthusiastic tennis players, perhaps the best in the regiment, and our sets were fiercely contested. White instructed me not to play a "customer's" game with him, that is, letting him win because he outranked me. I didn't let him win, but he did well against my best chops and spins and cuts. Those afternoons when his hard shots were missing and my slow stuff drove him crazy, he would occasionally lose his mild temper and disdainfully ask me when I was going to stop "chicken-shitting" around and hit the ball "like a man."

Field headquarters in the Bougainville jungles, G2 Lt. Col. Sears and the author

A number of Red Cross live shows came our way: Frances Langford, Randolph Scott, Jerry Colonna, Bob Hope, Jack Benny, Martha Tilton, Larry Adler, Carole Landis, and even Leo Durocher entertained the regiment and dined at our Regimental Officers' Mess. These celebrities socialized well, seemed more awed by our stories of battle than we were by their celebrity, and the soldiers loved the attention and the reminder that back-home had visited them at the front.

Chapter 10: Bougainville Mopup, and Ready for Philippines

Movie star Randolph Scott mixing with troops in the Solomon Islands.

But, from May to December, along with the fun and games came the most rigorous special training our troops had ever had. Close order drills, inspections of barracks, firing on the range were all resumed, to sharpen discipline and combat skills. In addition, we were brought up to date on the latest tank-infantry and amphibious assault tactics. We were readying for wide open plains and the congested cities of the Philippines, and it seemed that our jungle fighting days and nights were over. This was a happy realization, because now the fighting would enable us to maximize our firepower, artillery, mortars, and air strength instead of the primitive one-man-and-a-gun against one-man-and-a-gun.

We practiced loading and unloading a Landing Ship Tank (LST) which was to take some of us to Luzon and the Lingayen Gulf beaches of the Philippines. We practiced climbing up and down the nets on larger boats, and the highlight, or lowlight, of my loading practice came one afternoon when several of us took a motorboat to the troop ship.

Bill Leathers, the S2 (Intelligence Officer), had lent me a pistol for the occasion since he had an extra and I didn't want to be encumbered with a heavy rifle. We pulled up next to the ship in rather rough water. As the motorboat went up in the waves, White and the S2 jumped and hit the ladder entrance to the ship. I followed, only to be caught in the downswing of the motorboat and flung into the water, trying like hell to swim away from both boats so as not to be crushed in between their sides as they bumped each other.

The S2 Bill Leathers, a lanky, hard-bitten Texan, leaned over the landing net and put out his hands. Rescue was near. But as I lifted my hands to his, he screamed menacingly "You sonofabitch, you've ruined my pistol, hand it over."

And before he'd help me out of what, to me, threatened to be a watery grave, I had to give him the wet pistol, which he dried fondly with his right hand while he reluctantly used his left to pull me from an infantryman's most incongruous fate—drowning at sea!

The 148th Regimental Staff after the Bougainville campaign.
The author (top left) was adjutant at the time.

Chapter 10: Bougainville Mopup, and Ready for Philippines

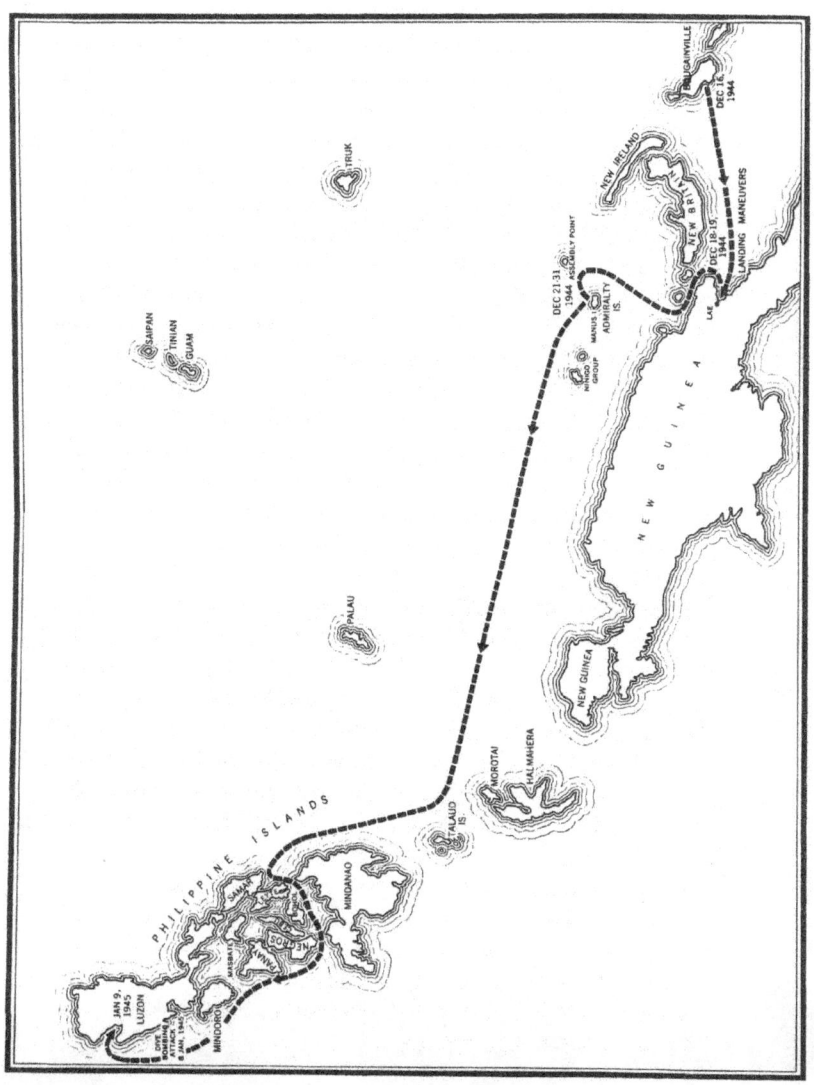

Journey from Bougainville to Luzon

Stanley A. Frankel

Chapter 11
The Luzon Beachhead

TRAINING WENT ON until December 14 when we all loaded onto whatever ship or boat we were assigned to and embarked for our final and fighting destination. We joined other craft as we steamed toward Lingayen Gulf in what was the largest single amphibious operation of the Pacific War, and probably second, in size only, to the Normandy invasion in Europe.

Regimental Headquarters and several companies were fortunate in being placed aboard a rather large troop ship where there was some elbow room, as well as adequate toilet facilities, fresh eggs, and beer. We sailed for three weeks, stopping once at New Caledonia to assemble with other ships, after circling to enable slower vessels to catch up.

The operation was no mystery to the Japanese who found many more targets than they had planes. In fact, our protection from accompanying carriers was excellent, and only a few Japanese kamikaze planes broke through to dive onto one of our ships. The panorama of hundreds of vessels of all shapes and sizes was awesome, and as we looked to the horizon where more and more ships kept appearing, we sensed that nothing could stop us and that the end of the war was surely in that flotilla.

Col. White briefed the staff and company commanders a few days before landing, a briefing which was informed, we thought, by intelligence from ashore via friendly Filipinos. White showed us the map, circled where each company was to hit, specifically x-ing out a small corner where my unit was to land and set up Regimental Headquarters. This time, I was going in with the first wave, and White pointed out that opposition would be sporadic, but fierce. The latest intelligence indicated that the Japanese 58th Independent Mixed Brigade had dug in near my unit's destination. This was the gang involved in the Rape of Nanking and White shifted a rifle company to my detachment because of the expected confrontation.

"Frankel" he said with a gleam in his eyes. "You will have the honor of commanding that part of our regiment which will have the hardest time."

I thought to myself that this was one honor I could do without, and I also made certain I carried a GI shovel with me so that the minute I hit the beach, I could start digging a foxhole to protect myself against enemy fire. I planned, if necessary, to dig all the 150 miles to Manila.

We circled the beaches the morning of January 9 while those beaches were being subjected to heavy naval bombardment. As I was climbing down the net from the large ship to the smaller Higgins (assault) boat which carried about thirty men, I slipped, but was caught by my supply sergeant, who quietly remarked that of course I wouldn't want to miss this invasion because of a broken leg. I thought to myself that I'd make that trade quickly. We started sailing toward land, but were advised that the Navy had another half hour of shelling before we would go in, so the assault boat circled. The water was rough; I had a queasy stomach, and in about ten minutes, I was seasick. Shells were whistling overhead and we thought that some of those shells were incoming from Japanese emplacements near the beaches. However, seasickness is one of the worst maladies I've ever had, and I got up from my squatting position in front, leaned my head over the side, and threw up.

The company commander in the boat screamed at me: "Frankel, get your goddamn head down or it's going to be knocked off."

I slowly reassumed my sitting position, looked at him, and retorted: "I hope so." I almost meant it.

Finally, we went in. The front section of the boat cranked down, and we charged the beach. The shelling was still intense, and I figured it was both ours and theirs, so as I hit dry land, I pulled out my shovel and began digging like hell. I honestly believed I set a foxhole digging record, for in less than a minute, I was almost underground, frantically throwing the wet sand all around the hole. Suddenly, I heard a strange noise as the bombardment quieted. This was the sound of human voices yelling "Veectorie."

I looked up, and to my astonishment I recognized the friendly faces of a dozen Filipinos who were then swarming all around us.

"Where are the Japanese?" I asked.

"All gone, two days ago, running to Manila."

Our intelligence, as usual, was flawed; there was no opposition to our landing and the only casualties I noted were a few dead horses, goats, and sheep.

It was about 150 miles from the beaches at Lingayen Gulf to the outskirts of our principal target, Manila. The Japanese were retreating as fast as they could, and we were running after them, as fast as our legs and jeeps and trucks could keep up. There were isolated incidents where, more by accident than design, the Japanese flight got caught between irregularly advancing American troops. These sporadic clashes slowed us down for a day, now and then, but the results were inexorably and monotonously the same: the Japanese were eliminated down to the last bleeding holdout, and the American troops moved in for their souvenirs, buried their dead, ate their rations, and kept churning ahead.

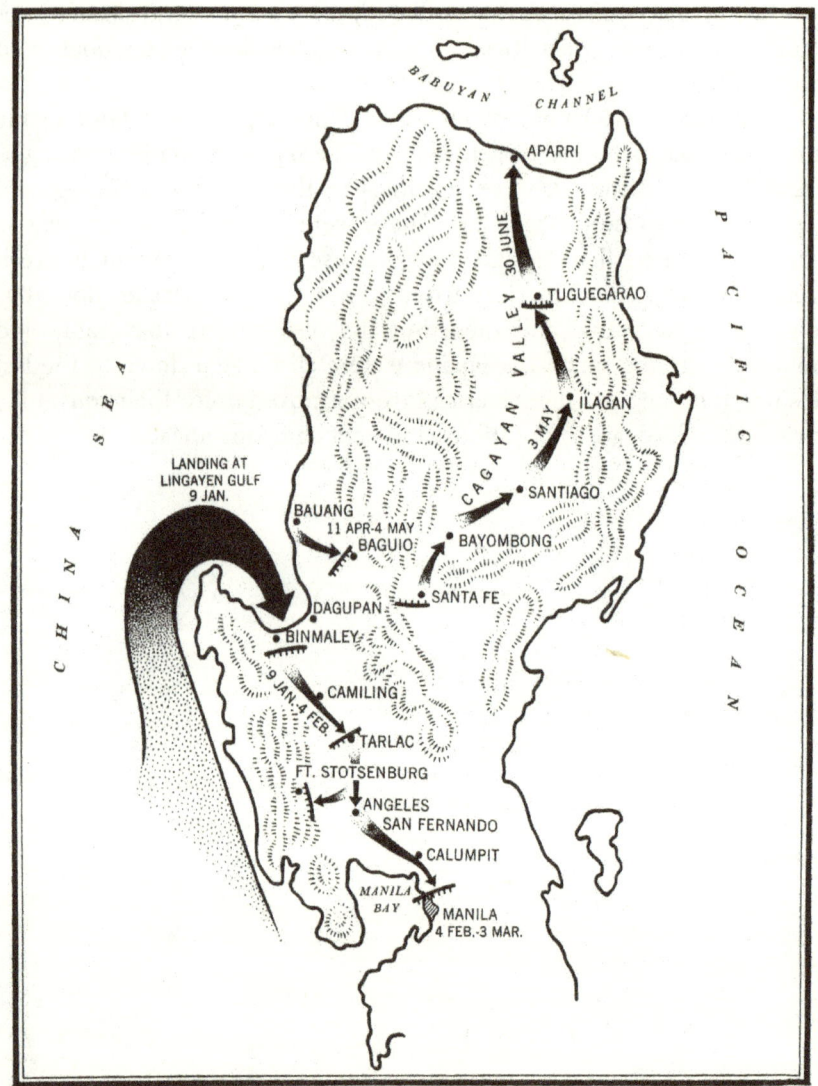

The Luzon Campaign

Chapter 11: The Luzon Beachhead

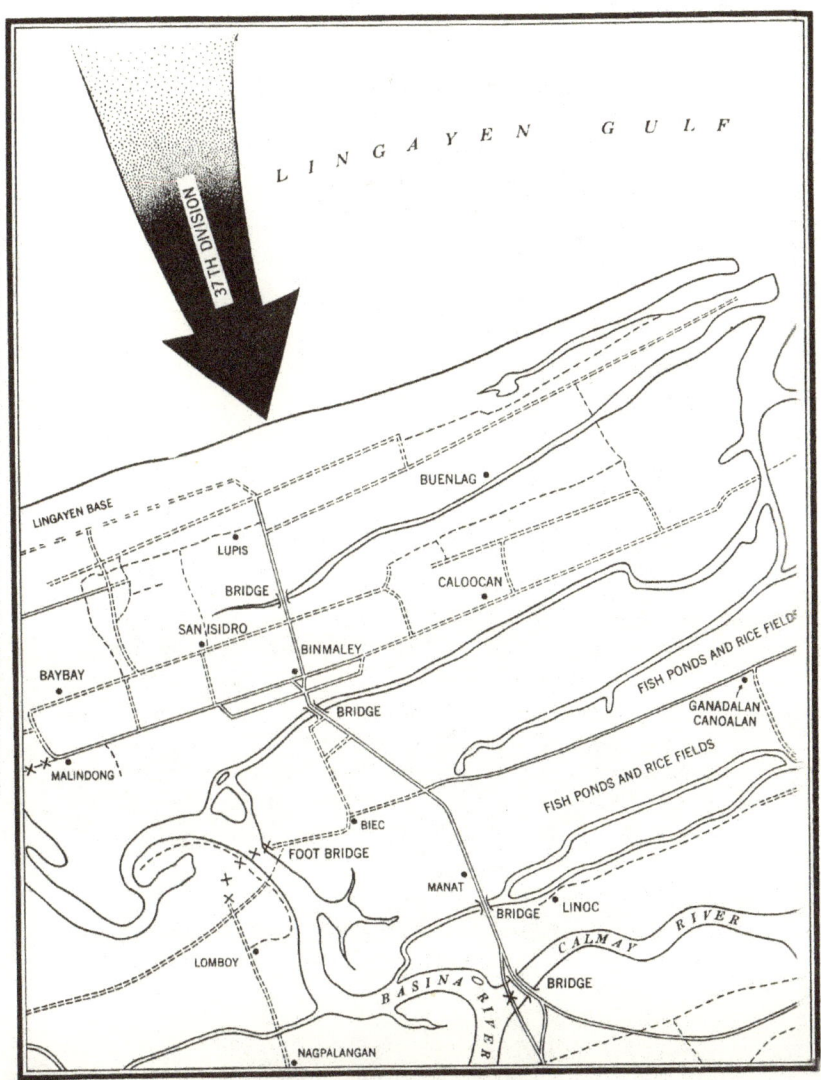

The Lingayen Beachhead

Stanley A. Frankel

Chapter 12
The Battle of Balintawak

IN THE GI'S BOOK, the "capture" of a brewery just this side of the last water barrier to the Manila suburbs may rank as the happiest few hours in the long combat record of the Regiment.

At 0930, February 4, the lead scout of the 2nd Battalion column, 148th Infantry Regiment, walked up to the gate of the brewery, and kicked it open. He saw warehouses, workshops, vats, and office buildings. Twenty-six days and 135 miles ago, the scout had gotten off an LVT on Lingayen Beach, primed and ready for the Japanese, and then had begun hiking south toward Manila. Spearheading the corps drive, he walked, walked, walked. His feet softened by one month aboard ship were full of blisters with blisters. Sporadic clashes with Japanese patrols sometimes slowed his pace.

A bitter three-day battle at Plaridel, twenty miles from Manila, stopped him temporarily. There the last Nip resistance before Manila was blocked out and the 1st Cavalry Division was permitted to speed on in their weapons carriers, jeeps, tanks, and armored scout cars.

The scout motioned the squad leader of the point squad to come up, and together they entered the brewery grounds. They noted broken furniture, vats, warehouses with something burning inside one of them. The whole place was strewn with broken window glass and beat-up desks and chairs. He peered inside a warehouse, then pushed back a sliding door, revealing a huge vat. Dirty brown liquid was pouring out of spouting faucets and from holes chopped in the vats by picks and hatchets. As the scout and the squad leader slushed into the ankle-deep liquid, they noted that the stuff had a yeasty smell. It was ice cold. The scout sniffed a bit more, then took off his helmet, removed the liner, scooped some of it up, and poured it over his grimy head. A few drops trickled down his cheeks and onto a tongue that was expectantly stuck out of his mouth. Sort of half afraid he might be right. He tasted *beer*.

"Cold beer," he yelped and the squad leader hit the floor, lapping the stuff. Sure enough, cold beer. Pouring out of all the vats, bursting out of the pipes, running out of faucets, shooting out of odd shaped holes. All over the damn place. And thousands of gallons of it. The electric cry of "cold beer" was telegraphed down the long column. The farther back the news went, the more it was accepted as emanating from the nearest latrine; in other words, just a gag. A grizzled noncom from Company H marching at the rear of the Battalion column, one quarter of a mile from the brewery, saw the BBB sign up ahead and decided that the sign had suggested cold beer to one of the jerks who had started the rumor. No. You didn't run across cold beer when you wanted it most. They'd probably run across the beer one of these days at Baguio amidst a flurry of snow.

One mile back, at a dilapidated schoolhouse, the regimental adjutant was setting up his field desk, when the morning report clerk related the story. The adjutant wryly remarked that he knew men who saw mirages like that under conditions not nearly so trying, and he wondered why the poor bastards didn't see eggs in their brew, too. At Division Headquarters two miles in the rear, a liaison captain with the G3 section boldly concluded that it was just a ruse to get rear echelon soldiers to go to the front lines where they would be greeted by the raucous raspberries of the infantrymen and maybe an unfriendly sniper bullet or two.

At Corps Headquarters, a helluva ways back, a one star general brushed off his neatly pressed denims, and cursed, "I guess the old man will want me to go up and investigate *this*, too. Christ, first it's the prison camp with four hundred beautiful Spanish girls going to bed with our assault troops. Then it's Gen. Yamashito waiting in San Fernando to surrender. Now, it's cold beer at Balintawak."

By this time, 538 men of the 2nd Battalion (total present-for-duty strength that day) swarmed over the brewery. A nondrinking engineer detachment had begun construction of a jeep bridge across the Tulihan. Battalion Col. Herbert Radcliffe, not one to countermand an act of God, told his men they could take a two-hour break, until the bridge was completed. If the men enjoyed the shade of Balintawak, that was OK with him, he remarked with tongue in cheek. "Just don't drink the polluted water."

The dirty, thirsty GIs were indulging in the soldier's dream. They swam in beer. Bathed in beer. Lapped it off the floors, filled up their helmets, canteens, mess gears, rations cans, jugs—even soaked their handkerchiefs with it. The beer was ice cold and it was green. But it was

Chapter 12: The Battle of Balintawak

beer and it was flowing and it was available to all. No rationing. No standing in line. No icing it up. Some guys just sat on the floor and drank. Others put their mouths against some of the small holes in the vats like calves on a teat and hung on beyond all reason or understanding.

Thank God the brew was mild. As it was, the men were comfortably lit, not dead drunk. They were more intoxicated with delight than with alcohol, but many started to talk about licking the whole damn Japanese army. The regimental commander arrived and chewed right and left until he was informed that the men couldn't go forward anyway until the bridge was completed. He then called the Battalion CO and told him he couldn't sanction this beer party but as long as he didn't know anything about it and as long as Radcliffe's men would be able to fight when the bridge was built, he wouldn't object. Radcliffe called his company commanders and passed on the instructions, which were heeded only because the beer was weak and the men were strong.

At 1500, Gen. Beightler, divisional commander, and Gen. Griswold, the corps commander, came forward to find out what was holding up the progress. They went directly to the bridge and noticed that it would take another hour. They inquired where in hell the troops were. A frightened platoon leader explained that the soldiers were gathered in an assembly area shaded from the sun so that they could get the maximum amount of rest. Griswold and Beightler moved to the assembly area and were immediately impressed by the enthusiasm, the singing, and the laughing voices within the walls. Griswold commended Beightler on the high *espirit* of the men after the long march down the plains, and Beightler (who knew his men) gave his thanks and got awfully suspicious. When the two generals entered the gates of the brewery, a beer-happy GI with a helmet full of beer in one hand and his rifle in the other stumbled onto the four stars. He immediately straightened up, dropped his precious weapon to the ground, shifted the beer gingerly from the right to the left hand, came up with a snappy salute and followed it up with a low bow.

The generals demanded to see the officers in charge, but the officers, grade captain and above, had discreetly gone on inspection tours. A second lieutenant was summoned, and he told the generals about the one-hour break and the ice cold beer and the long march and the happy men…and…by God, general, what would you do. Griswold *hrmphed* and Beightler *hrmphed* and they stomped out of the gate, telling the second lieutenant: "We can't sanction this sort of thing, but as long as we know nothing about it, Old Man Krueger can't burn us. Just be sure those men are in condition to fight when the bridge goes across."

One hour and hundreds of gallons of beer later, a relaxed, subdued, happy bunch of GIs lined up on the road, made sure that the water was out of their canteens and the cold beer was substituted. The lead scout started forward, sweating profusely, but still alert and cautious. He crossed the bridge and headed down the Rizal Avenue toward Bonifcaio Monument, which he made in a half hour. He perspired excessively, but the cheering crowds and the sight of beautiful girls and lovely suburban homes and imposing buildings made him forget the inevitable stomachache. At nightfall, the battalion bivouacked several miles from Bilibid Prison. The Battle of Balintawak was over. In about an hour, as dusk settled, the lead elements of the regiment reached the Tulihan River, the last nonhuman obstacle to be crossed before entering the suburban area of Manila. The Japanese had blown all the bridges across the river, but the U.S. Engineers had begun constructing footbridges, out of which, within another twenty-four hours, reinforced wider bridges would be built to accommodate jeeps and trucks and artillery.

I was in command of a Headquarters Company of about 100 men, and we received orders, along with the rifle companies, to dig in on our side of the river, get a good night's sleep, and then at dawn the next morning, cross the reinforced structures to launch our attack on Manila.

As we were digging our foxholes, I noticed an entourage of five shiny jeeps pulled up to the small area manned by my company. A tall, handsome gentleman, smoking a corncob pipe and wearing, not the prescribed helmet, but a fancier scrambled-egg officer's hat, jumped out of the second jeep and jogged forward, in my direction, followed by aides, journalists, and cameramen.

This was my first and last direct encounter with Gen. Douglas MacArthur, and to admit I was shell-shocked and starstruck was understatement.

He called out to my men: "Who's in command here?"

By the strangest of probabilities, at that moment, I was. So I walked up to the general, began saluting, and reported: "I'm, sir, in command, sir, in this sector, sir. Capt. Frankel, sir, Headquarters Company, sir, 148th Infantry Regiment, sir." Also, as I "sirred" him to death, I kept pumping away at my salute while I did observe out of the corner of my eye that the cameras were rolling.

In a quiet aside, the general whispered: "Stop saluting, soldier," which I did. But I was told later I then started to bow. In any event the

Chapter 12: The Battle of Balintawak

general, turning his best profile to the camera, inquired: "What are your orders, captain?"

I told him, still "sirring," that we were to dig in on this side of the river for the night and then launch our attack the next morning at dawn, crossing bridges which by then would accommodate heavier loads than just foot soldiers.

I watched the starry look in his eye as he started to speak, in theory, to me, but really, to the whirring cameras and microphones: "Captain, do you see those flames in the distance? The Japanese are burning down Manila, and they are going to burn and butcher the entire population, including my boys who've been dying in Bilibid Prison for three long years. If we wait until tomorrow morning, they will all be dead. We must enter Manila tonight and rescue those men and the rest of the population." He then came out with a strange statement which still rings in my ears: "Captain, if you'll go in there tonight, I'll go in with you."

As I stepped back and pondered that statement, I was aware it was for effect. Even if he wanted to go in with the front line troops, his aides would have physically hauled him back. I didn't know how to answer this question, which was rhetorical anyway, but I was saved by some scurrying on the left as Col. White, hearing about MacArthur's presence in regimental territory, had rushed over. The two greeted each other warmly. (As I have noted, White had been a lieutenant under MacArthur in the Philippines, before the war, and had been transferred to the States a few months prior to Pearl Harbor.) Those two then conferred over maps and I stood aside…far aside. The decision was made to go in that night, crossing on the footbridges and bringing with us only weapons which could be hand carried. We did have a few native canoes on which we hauled ammunition and some mortars and machine guns, but, by and large, it was a soldier and his gun walking in to fight what he had been informed were well dug-in, well-armed Japanese, prepared for a last-ditch and more-menacing-than-ever fight to the death.

Stanley A. Frankel

Chapter 13
The Rescue of Bilibid Prison

IN A FEW NATIVE CANOES and over a flimsy bridge constructed by GIs who were riflemen, not engineers, the 148th Infantry Regiment crossed the Tulihan River, the last water barrier before Manila, at dusk on the 5th of February, 1945. The first foot troops entered the city in the evening as Capt. Sidney Goodkin and Capt. Lawrence H. Homan led their Companies F and E, respectively, down Rizal Avenue. Cheering crowds with flashlights, lanterns, and burning candles broke the terrible tension of the men, a tension engendered by a grueling 150-mile march from Lingayen and those sporadic, bloody little scraps with the bewildered fragments of a withdrawing Nip army. The cheers and kisses and free drinks, and loud "Veek-torie, Joe," and the apparent absence of the enemy compensated for the dust, the blisters, the aches, and the sunburn. The liberation of Bilibid Prison was the highlight of the first night and ensuing day.

The liberation of Bilibid Prison itself was prosaic and undramatic. An advance element of the regiment entered the prison and found 1,200 of their fellow citizens. The rescue was accomplished in the face of an intelligence report that the Japanese had an ammunition dump at Bilibid which they planned to blow up when the American soldiers arrived. Here are the details: that first night, the troops, in route column, had moved about three miles into Manila. As they approached the corner of Espana and Rizal, the lead scouts of Company F were fired on by a Japanese machine gun in a well-constructed pillbox that dominated the street. These were the real opening shots in the bitter battle for Manila, and they temporarily stopped the advance. The number of Japanese was unknown as were their intentions. Our supporting weapons wouldn't be up for a day and time was not short, yet. Japanese riflemen began potting at the troops, now sitting along the street on the curb, as the ecstatic crowds brought them water, rum, eggs, cigars, and open hearts.

Goodkin, who had won the Distinguished Service Cross for leading his cited-by-the-president company against the Japanese on Bougainville, sent his third platoon to the left, to bypass the Japanese resistance and to find out what the hell there was on his exposed left flank. When they hadn't reported back after an hour, he became worried and wanted to go out by himself looking for them, but Col. Radcliffe told him to keep his combat boots on and to dispatch a patrol instead. Goodkin picked Tech. Sgt. Rayford Anderson of the second platoon and nine men. Anderson, whose wrinkled face and bald head belied his twenty-six years—he was a dead-ringer for Ernie Pyle—was briefed and started out with his men. His mission: to locate the platoon, presumed "lost."

He skirted the machine gun cleverly. He had learned both caution and small unit tactics in the jungles of New Georgia and Bougainville; street fighting was similar. He advanced a half mile on the left flank and still couldn't locate his objective. He queried numerous Filipinos who told him everything else except where his buddies were. They told him about the "damn Japanese," about their three-year vigil, and about their love for the Americans. Anderson looked at his nine men and wondered how many would survive the patrol. He wasn't sent out to win a Medal of Honor, just to locate thirty men. He thought of returning with this information, but asked his men what they thought and they unenthusiastically voted to keep trying to find the platoon.

Indian-fashion, the ten men kept close to the buildings, followed the shadows, and moved toward Bilibid. They finally reached a position opposite the main entrance of the large prison and didn't like the silence or the absence of cheering crowds; it presaged bad things to come. As they slowly approached the prison gate, they noticed two Japanese sentries lolling about, talking monkey-fashion, probably asking each other where was the Japanese air force, where was the Japanese navy, where was Gen. Yamashita, and when were they going to be rotated. Ten men drew ten beads and, on Anderson's command, fired. Two of our little yellow brothers were rotated to Shinto Heaven. The firing awakened a Japanese machine gunner up ahead whom Anderson couldn't locate, and who evidently couldn't locate Anderson, for the gun fired straight down the middle of the street, hitting only a few lampposts.

Anderson figured that the other way around the prison was now more inviting, and he again asked the men whether they wanted to continue. None of them was so damn keen about this exploration, but they felt morally obliged to keep on until their mission was accomplished. As Anderson remarked, "We didn't want to be chicken."

Chapter 13: The Rescue of Bilibid Prison

The patrol backed up a little, went to the rear of the prison. Anderson noticed a side door and he told Staff Sgt. John Smith, his second in command, to take Pfc. Ray Henry (a company clerk who had gotten bored with paper work) and Pfc. Joseph M. Dilks, and form a covering force around the door while he and the six remaining men forced an entrance. Sgt. Billy Fox, winner of the Silver Star and Purple Heart for Bougainville, was ordered to open the door gently. Gently didn't do the trick and Fox drawled that he might as well shoot the goddamn thing. So without asking questions, he pumped two M1 shots into the lock and smashed the door open with the butt of his rifle. He popped inside ahead of the rest and the other six followed him cautiously. The seven men found themselves in a large storage room, running about 200 yards alongside the prison, separated from the prison by a thick wall. A bit of reconnoitering and the GIs found only darkness, one big rat, and the uncovered dung of the Japanese sentry (probably), which reassured them that this was the same filthy animal whom they had burnt out of the jungles.

The wall of the corridor separating it from the prison was made of stone and had barred windows, rather low, spaced about ten yards apart. There was also a gate which was locked and couldn't be budged. The windows were boarded up so that no one in the corridor could look into the prison. Anderson told Pfc. Donald Ammon, to pry a board loose from the window and Ammon took out his bayonet and began prying quietly. The prying didn't do much, so Ammon, muttering to himself about the "sonofabitch better give," pried not so quietly, and he was able to pull a board back so that a man could look through and see five yards in either direction. The left and right five yards revealed nothing. Anderson peered further into the prison and saw an open courtyard, in the center of which were fifty people huddled together. They were conspicuously silent; Anderson learned later that they feared these rescuers were really Japanese who had returned to finish them off.

Pfc. Robert Cernoch and Pfc. Marvin Fraikes (killed two months later at Baguio) called out loudly for someone to let them in. The fifty people didn't move, and as the men could see more clearly, they were certain now that these fifty people were white. The whole patrol then tried to persuade someone to open the gate, and Pfc. John Lamb even sang out a few bars of God Bless America, but to no avail. The people inside could make out only the silhouette of these jabbering men, and they didn't know about the new American helmet. They expected World

War I models, and they had concluded that this was just a ruse used by the Japanese to gain entrance.

Fox thought of this, had the men remove their helmets, told them to stop squawking, and began to talk quietly and earnestly to the people inside. Sgt. Smith tossed some Philip Morris cigarettes into the courtyard, and one brave old man moved forward, picked up the cigarettes, stared anxiously and fearfully into the faces ten yards from him and then hollered, "By Jesus, it's the Yanks!"

The others closed in, took a few skeptical looks, then began laughing and cheering, and finally opened the gates. The American internees deluged the men with sobs, hugs, and a multitude of questions, but the breathtaking occasion didn't make Anderson neglect self-preservation. He told Pvt. Ammon to go like hell back to Goodkin with full details and to ask for further instructions. Meanwhile, the rest of them would stay at Bilibid.

Ammon reached Goodkin and Lt. Col. Radcliffe, told his story, and Radcliffe immediately started his troops moving forward. The machine gun obstacle had been cleared, and Radcliffe had planned to keep his lines right there. Now, he was extending those lines to include the prison proper, so that the people inside would be under full protection of the American troops.

The battalion moved in and around the prison, relieved the patrol, and discovered there were over 1,200 military prisoners and civilian internees waiting desperately for the Americans. The full Japanese garrison had pulled out early that morning, warning the prisoners they would soon be back when the Japanese counterattacked and annihilated "Gen. MacArthur's bastards." The military prisoners were in pitiable condition; almost 200 of them were in the dingy hospital, many without arms, without legs, all without an ounce of surplus flesh. The civilian prisoners were lean also, but they were in much better nourished physical condition than the military—a premeditated Japanese policy.

Col. Radcliffe talked to some of the leaders, pieced their stories together, and called back to Regimental Headquarters by the phone which had followed him up the line. He told Col. White, "We've got a city of white people on our hands. Please send us food, trucks to get them out of here and medical supplies and doctors."

Food, medical supplies, and doctors were sent and more came the next day, but the present motorized equipment of the 148th "flying columns" were three jeeps, one ambulance, and a few caribou carts. The footbridge over the Tulihan had served the infantry well, but the

Chapter 13: The Rescue of Bilibid Prison

motorized columns had to wait for the overworked engineers and their Bailey bridges.

Anderson told his Company Executive Officer, 1st Lt. Oliver Draine, holder of two Silver Stars from New Georgia and Bougainville, "That hundred-and-twenty-five-mile march down from Lingayen was awful tough, but by God, lieutenant, the look on the faces of these people paid me back for every damn blister. I'll never forget that look in their eyes. Did you locate the lost platoon?"

They had.

The second phase of Bilibid's liberation, just getting out the 1,200 American prisoners, was more ticklish, though also without much danger.

The prison's southern extremity was within 200 yards of the Pasig River. The Japanese across the Pasig River, in great strength (15,000 soldiers and sailors), had begun their demolition tactics on the 4th of February. The heart of Manila was going up in flames and explosions. Japanese rearguard troops on the American side of the Pasig slowed our troops, and blown bridges delayed the supporting armor and equipment. On the next day, with the Japanese torching gasoline drums and blowing up buildings as fast as their little brown hands could push detonating buttons, the situation darkened. The fires, fanned by a strong wind, began to jump the Pasig at a point near the prison. Great amounts of debris started dropping onto the prison grounds, thrown up by the explosions across the river, and several of the leaders of the prisoners excitedly informed Col. Radcliffe that a Japanese storehouse of ammunition and fuel, located near the prison, would probably be set off by the Japanese if the flames did not blow up the dump first. Higher headquarters were notified of the imminent danger.

At 6:00 p.m., Gen. Charles Craig, assistant division commander of the 37th Division, came down to inspect the predicament. He quickly grasped the growing hazards and possible consequences to those now at the prison. He ordered Col. White to begin evacuating the prisoners and he promised that he would dig up trucks for those unable to walk. He returned to Division Headquarters where he began to work his magic. Every unit within driving radius was summoned, and he was told that enough transportation would get to Bilibid…given enough time. However, the fire and the Japanese ammunition dump might be a bit reluctant to wait for the trucks.

Col. White detailed three of his staff to supervise the evacuation and arrange the details. In charge was Lt. Col. Delbert E. Schultz, Maj. John J.

Gallen, the regimental surgeon, and the adjutant. Col. Radcliffe was responsible for the overall security of the evacuation, and he deployed his battalion around the prison in blunt disregard of the dangers of Japanese and flames. Hundreds of Japanese, caught in their own handiwork, were being driven back toward our lines and in the direction of Bilibid. Their scorched pants didn't cool their fanaticism, and they fought bitterly, if in vain. Radcliffe was unable to spare personnel to help in the evacuation since his lines were already thin.

Col. Schultz and his two assistants, together with a motley assortment of trucks, medical aid men, litter bearers, clerks, communications personnel, and any strays who were not committed to the fighting, arrived at Bilibid at 7:00 p.m. The trucks now on hand would haul about 100 ambulance cases. At least three times as many were required. The flames began licking at the gasoline storehouse on Ascarraga Street, immediately south of Bilibid, and the rescuers were frantically organizing a speedy system of evacuation. Col. Schultz divided the internees into the walking and non-walking, being assisted by Doc Gallen, who had to be ruthless in his sorting. All who could walk, must walk. The adjutant was detailed to organize the evacuation of the 300 non-walking cases, and he tore his already-sparse hair trying to figure out how to get the hospitalized onto the waiting trucks. Utilizing Filipino sightseers, the strong walking internees with more guts than muscle, truck drivers, litter bearers, and even a platoon of infantry from Col. Radcliffe's hard pressed men, he got the loading underway.

The walking internees, once oriented and shoved, began moving fast up Rizal Avenue. Crowds cheered these men, women and kids; snipers took a few shots at them. Driven inexorably by crimson-faced Col. Schultz, the 900 walking prisoners toddled slowly toward safety. The litter cases were still being carried carefully through the one small doorway and loaded exasperatingly slowly. There weren't enough litter bearers and the patients didn't seem to realize the urgency. They bitterly resented being hurried. One woman insisted on taking her mattress with her and wouldn't leave without it. Another old man, who tottered pitiably, refused to be assisted, saying he walked into the prison, and he would, by God, walk out. Their spirit was inspiring, but it was hard on the blood pressure of the American troops. Trucks finally began to roll in from all units, and the 1st Cavalry Division radioed that it was sending a large fleet of 6x6s.

In the meantime, the 2nd Battalion was engaging in a dozen little firefights as Japanese would pop up inside buildings, behind our own

Chapter 13: The Rescue of Bilibid Prison

emplacements, and from nearby sewers. The Japanese were a heterogeneous, ragged, confused crew, but they died hard. The infantrymen sensed the greater danger of the oncoming fires and took on the Japanese in a business-like, casual manner. Not one GI was panicked by the potential destruction advancing toward them. They realized that they were holding the line to protect 1,200 men, women, and children. It would have been senseless to maintain this front otherwise. Now, it was unthinkable to do anything else.

Company E had an observation post in a small tower in the middle of the prison. The soldier at his post spotted Japanese movements and using a sound power phone, called down locations to a mortar crew who kept a barrage of 60-millimeter mortar shells dropping on unwary enemy in any group larger than one. At 10:00 p.m., the observer in the tower was located by the Japanese, and they directed devastating machine gun fire at the tower. The observer was badly hit, and he called out on his phone that they had "got" him. Pfc. Elmer Russell, an aid man, and impish-looking Chaplain Elmer Heindl, didn't wait to volunteer but just started up the wooden stairs to the tower.

Another long burst of machine gun fire ripped the tower and the wounded man was hit again. He had only a few minutes to live. Russell and Heindl ran into the tower and began aiding the dying soldier, Russell binding his wounds and the chaplain giving him last rites. Russell lit his flashlight for an instant to determine where the man was hit most severely, and when the light was flashed on, the Japanese again opened up but miraculously missed the three men. When the observer died, the Chaplain and the aid man carried his body down the steps to the prison yard. For this heroism, both men received the Distinguished Service Cross, Russell's arriving posthumously since he was killed a week later while engaged in a similar mission.

Two and a half hours later, more trucks arrived, and the truck drivers were mobilized as litter bearers. The desperate speed with which all of the litter bearers now worked was inspired by their desire to get the job done and get the hell out of there. The last 100 patients were finally loaded and an inspection was made of the hospital to make sure none had been overlooked. Satisfied, the adjutant plopped himself on the front end of a jeep and led the convoy up Rizal Avenue. Crouching low and emulating a radiator cap statue, he proceeded by Rizal, acknowledging neither the wild bullets nor the wild cheers.

The trucks were still pouring in, and they now (like Fifth Avenue buses) began to pick up the walking internees. In another half hour, all of the civilians had been entrucked and were on their way to safety. Only then did the GIs begin to grumble; Col. Schultz was a bit uneasy himself. He made a flying trip around the prison and found only soldiers with long faces. He had several of the empty trucks back up against the warehouse behind the prison to pick up cots and foot lockers belonging to the prisoners, a bit of foresight which endeared him to the internees but which met with little acclamation from the impatient GIs. Finally, when the cracking of gasoline drums and exploding of small arms ammunition warned the group that it was past time to depart, Col. Schultz ordered the troops to withdraw, which they did in a hasty though orderly manner. As the rearguard pulled out, the Japanese riflemen and snipers fired wildly at the withdrawing targets, but the GIs escaped with only a few minor casualties.

Satisfied that none remained between him and the Pasig except Japanese with hot britches, Col. Schultz got into his jeep and told his driver, Whitey Henderson, to take off. A couple of farewell shots zipped by the windshield, one bullet nicking the rear vision mirror, but Schultz, in a final measure of disdain, turned around, thumbed his nose at the prison area and then asked Henderson if he were driving a carabou [sic] cart or an automobile and to step on it.

The internees were deposited at a shoe factory six miles away from the prison, and they were feted with K rations and eager listeners. The wind changed later on and the fire didn't engulf the prison as expected. Only a small section of the northern part was singed, and, contrary to reports, no ammunition dumps were discovered.

Two days later, when the fire had completely subsided and the prison was safely behind our lines instead of in the middle of them, the internees were taken back there to await transportation to the United States. In a few more days, the first of many prisoners said goodbye to Bilibid for the second and last time and were homeward bound.

Chapter 13: The Rescue of Bilibid Prison

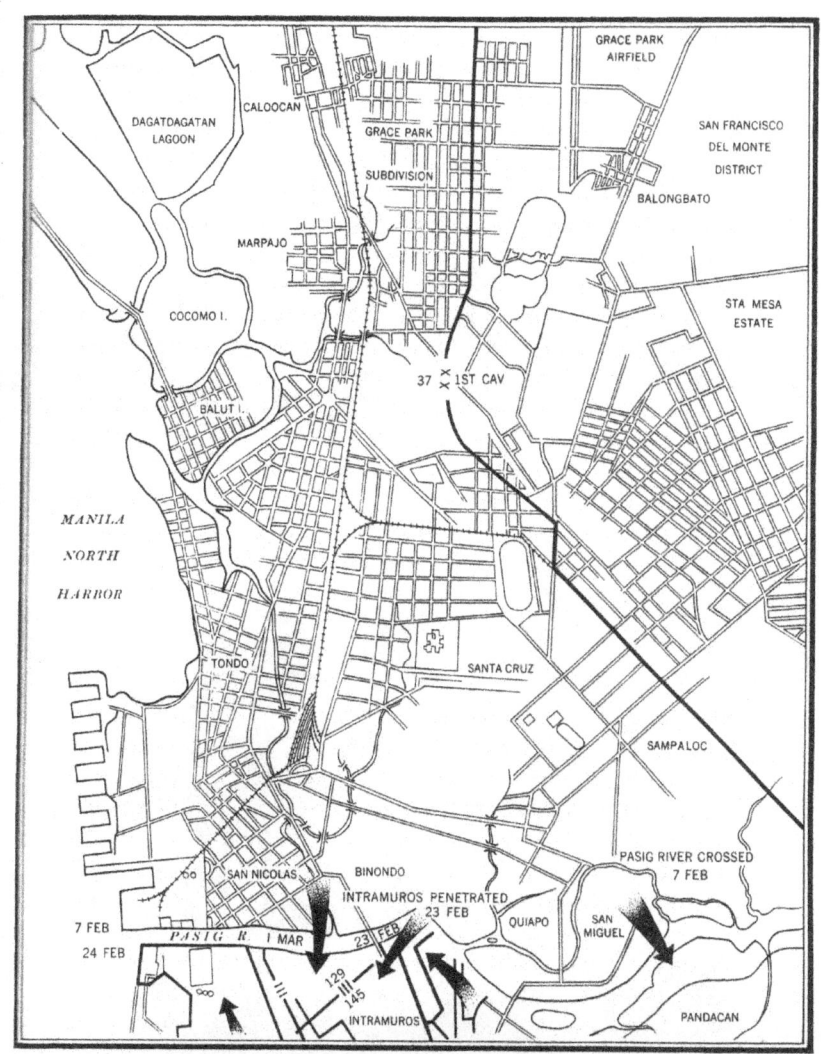

Manila North of the Pasig

Stanley A. Frankel

Chapter 14
Incidents in the Battle for Manila

THE BATTLE FOR MANILA lasted for the month of February, 1945, and into March. Accompanying the heavy fighting were many small-scale incidents which served, in microcosm, to tell the story of the elimination of the Japanese forces in Manila. These bits and pieces may be classified as mere footnotes. But what footnotes!

Pfc. Cletus Rodriguez and fifteen men made the first floor of the Legislative Building. Five of their comrades were strewn around the entranceway, cut down by Nip machine gun fire. As they entered the waiting room, fire from snipers and light machine guns came at them from all the corridors and especially from the winding stairs directly in front of them. Rodriguez knew they would all be killed if they stayed in this open spot, so he jumped up from his prone position next to the stone statue of Jose Rizal, pointed his BAR forward, and yelled "follow me." He dashed up those winding stairs as the Nips began throwing the kitchen sink at him. He kept his finger down on the BAR trigger, pointing it this way and that, up and down, right and left, at any and all obstacles in his race for the third floor. He was hit a few times, lightly, but he kept firing and racing upstairs as the dead enemy sprawled in front of him, fell over the banisters, or rolled down to the bottom. He reached the top, eliminated the final machine gun crew, which had commanded the stairway by shooting the three gunners in the back, and then, things now quiet, he looked behind him for the men he believed had been giving him close support the whole way. Cletus Rodriguez found himself alone. The men just hadn't started when he had, and were still hugging the first floor crevices and debris.

Rodriguez was really scared then, and he dashed down the steps screaming at the top of his lungs: "You sonsofbitches. You killed me!" Sobbing, he was mobbed by his comrades, and he followed them around, sheepish and afraid, as they mopped up the rest of the Nips in that sector of the building.

Our side had finally gotten a foothold, and two days later, after fierce and bloody fighting, the whole building would be ours. For his heroism, literally single handedly, Rodriguez was awarded the Congressional Medal of Honor, one of the few going to men who had survived their bravery.

Then, there was Tech. Sgt. Wren, who heard funny noises in the basement of a house that had been knocked down by our own direct artillery fire. The only entrance left to this basement was a slight hole through which he could squeeze his body. He didn't know how far a drop it was from the ground to the basement floor, so he grabbed a nearby mattress, jammed it into the hole, and heard it land at the bottom. Moving his carbine off safety he jumped down into the basement, alighting gingerly on the mattress. It was hell-dark and Wren could barely make out the silhouette of a man. He called and the man jabbered back excitedly. Wren took careful aim and shot him. That really ignited things, for the basement suddenly came blazingly alive. From all sides there was firing and screaming and grenade throwing. Wren backed into the farthest corner, and added to the din by firing rapidly at every noise or form he could distinguish.

After twenty minutes, during which he had emptied sixty rounds in the basement and tossed five grenades into the corners farthest from where he had ducked behind iron pipes, things quieted down. He noticed one lone figure trying to crawl out of the same hole from which he had entered. He shot the figure in the buttocks, and when it dropped down onto that mattress, he shot it in the head.

A few minutes later a cautious Yank called down the hole: "For Christ sakes, Wren, are you OK?"

"Sure," replied Wren. "Where the hell were you brave guys when the fighting was going on? Bring a flashlight, and we'll see what we've got."

The soldier came down, sans flashlight, lit a match, and whistled. Strewn about the large basement were the bloody, torn bodies of sixteen Japanese, who had thought a major attack was on…and who had shot hell out of each other in a vain effort to get one cool Yank ducking in a corner.

Wren crawled out of the hole. "Good to get a breath of fresh air," he drawled. "Smelled to high heaven in that damn basement."

When the fighting had died down and the cleanup started, our men were now more interested in souvenirs than Japs.

Chapter 14: Incidents in the Battle for Manila

An old woman ran up to two of our soldiers who had been hunting those souvenirs along the battle-scarred streets. "Japs there, Japs there," she pointed.

The boys were tired and they had heard this one before. "If there are Japs, you come show them to us."

The old woman, scared to death, stealthily led them to a large piece of tin roofing covering a small hole. "There," she whispered.

One of the soldiers quietly lifted up the tin, and the other, unconcerned, pointed his carbine into the hole, which he assumed was empty. The tin came up, and two pairs of eyes gleamed ominously from the hole. Jim shot twice; John dropped the tin roof. Both started walking away, until Jim turned around as an afterthought: "Thank you, ma'am," he said, and went back to the more important business of finding souvenirs for his gal.

Col. Delbert Schultz, who replaced Col. White as regimental commanding officer, was hard-nosed, rough, ill-tempered, but with an uncanny ability to get things done—to bull them through. I accompanied him on many of his jaunts, but fortunately missed the ones that resulted in the greatest danger. He had an explosive temper, and once, while returning from the front lines in his jeep, a sniper took a potshot at him, hitting the windshield of his car. The firing was coming from the second floor of a house nearby, and, completely losing his cool, Schultz charged up the stairs of the house, even forgetting to take his pistol with him. He ran into the room and saw the Jap, a husky sailor, leaning against the window, watching for more meat. Schultz jumped him and the big sailor put up a struggle, but Col. Schultz knocked him down, then hit him over the head with the Nip's own weapon…and brought him back to camp. En route, the Nip tried to jump out, but Col. Schultz cracked him one again. Quite an experience, and typical of him when he was drinking heavily or got his dander up, which was often.

When I had been promoted to regimental S-1 (adjutant) at the end of the Bougainville campaign, I had been assigned an assistant, Warrant Officer Sayre Shulter, the first professional male secretary I'd ever met, one of the kindest, gentlest, most efficient men in the Division. From a letter to Irene, here was my description of him:

> Thirty years old, married, he came up through the ranks to his present administrative job. He is by far the finest shorthand man and typist and the most meticulous person I have ever met. He is quiet, shy to a fault and very modest. Knows his administration

cold, and he will do any job I give him in a superior manner. Anything. Had some journalism, went to a business college, and his mind is keen as a bayonet. Little fellow, very thin and wiry. Loyal, and never takes any credit for anything and since I am responsible for everything that comes out of this office, he makes me look very good at times. His only problem is that he cannot delegate work. If I give him a job to do that is too big for one man, he won't think of calling on some of our very capable enlisted men to help him. Instead, he'll stay up half the night and do it himself. Which is tough on him, and something I am trying to break him of. He and I should be able to do everything in this office but we should merely supervise the work and let the excellent and plentiful non-coms and technicians take care of the details.

Shortly after my letter to Irene, I had the unhappy task of writing twice to Sayre's wife. The first was a letter advising her of Sayre's death and informing her of the posthumous Silver Star Award honoring him for his gallantry in the Philippine action. She answered, asking me some question about his records, and I responded as fully as I could. Between the lines, which I never revealed completely, were the following circumstances: Sayre and I were working in our Malacañan Palace Regimental Headquarters, readying orders for an attack by the regiment the next day across the Pasig River which bisected Manila. It was to be our final drive against the 25,000 Japanese crammed in the lower half of Manila backed up against the Pacific Ocean.

Captain Frankel in Manila briefing his staff
on plans for crossing the Pasig River the next day

Chapter 14: Incidents in the Battle for Manila

As we were writing the battle plans, one of our soldiers ran into our small workroom, screaming: "Lots of guys have been hit by Japanese shelling in the courtyard. We need litter bearers to get them back to the aid station."

Sayre jumped up, running toward the door to the courtyard, and I followed, unenthusiastically. The shelling had subsided and he and I helped a few of the wounded from the yard to the aid station. We went outside again to make sure that we had gotten all the wounded. The shelling resumed and an explosion near us knocked Sayre and me to the ground.

I was only shaken up, but he yelled: "I'm hit."

I ran over to him, helped him to his feet, and did not detect any visible signs of a wound. I put my arms around him and asked if he could walk to the aid station.

He replied calmly. "I can run."

Sayre, partly supported by me, and I jogged toward the aid station, about 100 yards away. I assumed he was in satisfactory shape, but as we entered the aid room, he broke away from me, dived toward the surgeon on duty, and lay there, flat on his face.

The doctor hurried over to him, turned him on his back, felt his pulse, and said quietly to me, "This man is dead."

"He can't be," I yelled somewhat hysterically, whereupon the doctor opened Shulter's closed eyes, felt under his neck and then his wrist, and confirmed:

"Dead. Dead. Dead."

Later, I was told that the concussion had killed him, but a dying impulse had apparently kept him going for the few moments it took to reach the doctor...then he had collapsed and died.

His official title had been Assistant S-1, Chief Clerk. He was killed 200 yards from his typewriter.

Finally, on a lighter note: almost a year after the war had ended, I read in a Chicago newspaper a story announcing that Dr. Jose Laurel, puppet premier of the Philippines during the Japanese occupation, had been granted a thirty day leave from the prison where he was being held awaiting trial for collaboration so that he might visit his home in Manila, Malacañan Palace.

On reading the story, I wondered what he might find when he returned to the palace. I knew it well. During the early stages of the battle for Manila, before the fighting crossed the Pasig River, it has been the command post of our regiment.

A sturdy, massive stucco building, the palace had withstood the shelling of the previous days and when reached by our troops still housed some of Laurel's palace guards and a few hangers-on relatives.

Both the guards and the relatives were passed through our lines to the rear, and the reluctance of the guards to leave their job was quickly dissipated by a few dirty looks. Our troops were in no mood to bargain. Once inside the palace, systematic operational planning and systematic looting proceeded apace.

First spied was a large, brown safe. Outside the safe were stacks of Japanese occupation money, which revealed Laurel's state of mind. When the engineers blew open the safe they found 40,000 pesos in good, old Filipino money—the kind backed by American gold—which Laurel obviously valued a bit more than his Japanese Mickey Mouse paper.

The counter intelligence boss took over the currency, but from then on, it was every man for himself. No warm meal was on the fire when our troops arrived, but everything else was pretty much undisturbed. Hundreds of beautiful dresses, children's toys, shoes, bottles of perfume, and trinkets were picked up by the searchers. Later, Filipino girl friends paraded around Rizal Street in dresses which had last been viewed at presidential balls, and a lucky Filipino man strutted around Fortuna Street in the top hat and tails which Laurel had worn to his inauguration.

In another room we spied a stack of new books. Fresh off the press, called "Man of Destiny," written by a Filipino writer who was of little honesty and much hunger. Laurel was extolled as the savior of the Filipino people, and fascinating pictures dotted the chapters of our boy, in palsy poses with Gen. Homma and other slant-eyed notables. Laurel's Declaration of War against the U.S. was hailed as brilliant, and the Japanese were publicly thanked by Laurel for bringing to his people the freedom which America had promised and postponed for so long.

In all fairness to Laurel, several photographs mentioned that he had stood up to the Japanese in their demand that the Filipinos be drafted into an Axis army.

"No," said Laurel, "we do better by working on the farms."

Later when intrafactional disputes and Filipino military inexperience destroyed the GI's faith in the Filipino's fighting ability, Laurel was proved correct. Better stay on the farms.

His study was plastered with diplomas and degrees, and one we recalled particularly was from Columbia University, New York. A photo album showed him with American senators and American generals, with Laurel wearing the same grim smile with which he welcomed Homma. In

Chapter 14: Incidents in the Battle for Manila

short, Laurel was ecumenical. A stamp album of American and Asiatic stamps was readily claimed by a philatelist carrying a tommy gun in his right hand. Some of the stationery with his impressive imprint found its way into the homes of the Smyzikis from Chicago and the Goldbergs from Brooklyn.

A large bin next to the study was packed full of bags of rice and cans of salmon and sardines, a bit unworthy of a president but indicative of Laurel's realism. People were starving and he naturally laid in a store of food. The rice and salmon were soon passed out among the Filipinos who grabbed it happily, and the sardines spiced up the meat and vegetable stew of our own GIs.

Altogether great quantities of cheap jewelry, furniture, pictures, vases, typewriters, electric light bulbs, combs, chairs, and even a couple of radios and Frigidaires were spirited to the rear lines.

The most fun among all the denuding was the wall tapping. One evening, a special service officer with little to do during the fighting began shadow boxing the walled panels. A couple of good knocks on one panel resulted in its opening up. He immediately reached in and came out with a quart of Canadian Club. Further excavation in that treasure chest brought forth more liquor, boxes of Corona cigars, some tinned caviar, and Del Monte peaches. That find initiated an intense wall-knocking search, but we must admit that some of the boys were not so subtle. When they suspected the walls of treasuring more scotch, or more cigars, they didn't dally long with the light tap of their knuckles; they just butted the hell out of the wall with the stock of their rifles. Much of the prospecting was fruitless, but some of it brought forth riches, not in gold or silver, but in liquor and cigars, which at that stage of uncertainty meant more to a GI than a bank-full of dollar bills.

After our regiment vacated Laurel's home to move across the Pasig River, a rear echelon headquarters moved in. I never heard what happened to it after that, but for our regimental officers, it had provided an interesting change from our typical field headquarters.

Stanley A. Frankel

An American observation post in the City Bank Building, Manila

Chapter 14: Incidents in the Battle for Manila

Portraits of Manila after its Capture

Great Eastern Hotel

City Hall

Stanley A. Frankel

Chapter 15
The Damn Fool

The manila battle, for which my Regiment received the Presidential Citation (awarded to army units conducting themselves in such a manner that, were they individual soldiers, would earn them a Distinguished Service Cross) produced many heroes. Four 148th infantrymen won the Congressional Medal of Honor, among them one of my friends, Bob Viale, whose actions—and death—occurred not too many blocks from where our Regimental Headquarters were temporarily set up:

THIRTY SECONDS elapsed between the explosion of the hand grenade and Lt. Viale's death. We don't know what he thought about in those thirty seconds, although we heard him gasping, "Damn fool, damn fool," over and over again until he died. His intestines were so badly ripped and splattered by the grenade that probably he wasn't able to think at all.

Whether he actually knew what he was saying or not, the rest of us acknowledged that Bob Viale was a damn fool. Even when he reported to us at Bougainville, fresh from Fort Benning and the States, we suspected his slight lisp and naive earnestness. We assigned him to Co. K, the reserve company of the reserve battalion, and his platoon leading consisted of heroic efforts to erect homes out of shelter halves and to cut latrine seats out of oil drums. By the time Company K was sent into the line, the Japs had expended themselves completely and were now engaged in foraging rather than fighting. We didn't wander too far out of our perimeter to hunt them down, and they didn't come too close to our fields of fire to dig edible plant life. Viale never fired a shot in anger the whole six months he spent on those northern Solomon isles.

Early in December, we boarded transports to make the grand assault on Luzon, Philippines. The brass hats briefed their staff meticulously, and even the platoon leaders were given all details. In turn, the platoon leaders, at special sessions on deck, carefully outlined the big and little picture to their men. Each man learned the army mission and the squad mission. That was standard operating procedure. Viale carried this orientation to a ridiculous extreme. Each afternoon he'd summon his platoon sergeant, platoon guide, and squad leaders to the officers' mess of the USS Harris and there plan grand strategy—on platoon scale. With diagrams, maps, field manuals, notebooks, rulers, pencils, aerial photos, and what not, Viale and his "staff" would prescribe each tactic and maneuver. We joked that Viale had even deduced the exact minute each man should urinate, and we explained to him that "the best laid plans of mice and men" are often shot to hell if a guy has kidney trouble.

He won the nickname "General," and some of his fellow officers conceived the cynical notion that since the regimental commander was on the boat, Viale was showing off. The colonel did notice Viale's enthusiastic briefing and "encouraged" other platoon leaders to emulate the "General." Which, of course, further endeared him to his comrades.

We soon discovered that Viale's fanatical concern for the welfare of his platoon was not just eyewash. He spent more time in the crowded holds, visiting his boys, than he did playing cards or reading in the comfortable officers' mess. He begged extra candy and cigarettes for his men, personally escorted them to sick call, gave them stationery and stamps, snuck them up to his quarters for a shower, and even shared with them a bottle of Three Feathers Bourbon which he had salvaged from Officer Liquor Distribution Days at Bougainville. This last gesture, everyone remarked, cinched it. He wasn't an exhibitionist. He was a damn fool. That was Bob. His sympathetic attention to his men shamed other platoon leaders. Although they cursed Viale bitterly for his "un-officer-like fraternization," they soon had their own quarters looking like a noncom's club. Nor did the brass discourage this friendliness. These were the men you would order to die for you in a month, and the platoon leaders were given complete freedom to work out their own brand of discipline.

At 9:30 a.m., 9 January 1945, the Luzon assault began. The rehearsals and the skull practices paid off, as each squad went for its predesignated palm tree, church steeple, or bridge. The Japanese disappeared and the cost to our regiment of establishing this beachhead were one man gored by an enraged caribou, and two men hurt as they fell from the landing nets into the LCT's. Then followed the grueling 137-mile hike down the central plains, toward Manila, the men limping along on blisters with blisters. The Nip resistance was spotty. There were small scraps at Clark Field, Fort Stotsenburg, Plaridel, and Malolos. There were patrol clashes

Chapter 15: The Damn Fool

on the flanks. But the longest stretches were nothing more lethal than prostitutes charging three cigarettes or two Atabrine tablets. On 4 February our troops crossed the Tulihan River on a bridge impoverished with oil drum pontoons and moved cautiously into the outskirts of the capital city Manila. As the troops talked gingerly down Rizal Avenue, they were deluged with cheers, liquor, and kisses, the hullabaloo drowning out the occasional ping of a sniper's bullet which presaged ominous things to come.

On the morning of 5 February, the 3rd Battalion, 148th Infantry Regiment, in the Bisondo District, ran into stiff Japanese rear guard, which had to be ferreted out of churches, restaurants, and storehouses. The Nip had traded his jungle for the wall of a store and his tree and vine for a second-story bedroom window, but he hadn't lost the sneak-touch. The battalion objective this day was the Jones Bridge across the Pasig River, and the Nips' objective was to hold the north bank until the bridge could be destroyed and the heart of the capital, across the Pasig, could be gutted, burned, demolished, blasted by fire, and explosives. Five hundred yards from the Pasig was a small stream running parallel to it, the Estero de la Reine. Co. K was pacing the battalion attack and Viale's platoon, the first, was the company's point, way out in front.

By 11:00, the flames across the Pasig jumped the river around Jones Bridge, and a strong southeasterly breeze fanned the fire to the right and front of the attacking elements. It was necessary to shift the direction of the attack to the left and to bypass the smoldering sections if possible. Viale angled his thirty-five men toward the fifteen-foot-long Ongpin Street Bridge, crossing the Estere de la Reine. As the lead scout darted across this bridge, he was halted by crossfire from three different pillboxes across the stream, commanding the span and its surroundings. The scout squirmed back to Viale and reported the exact location of the three pillboxes. Good boy.

Without hesitation, Viale formed a three-man assault team, consisting of a veteran squad leader, a BAR man, and himself. He ordered his rifle grenadier to toss a few smoke grenades on the opposite end of the causeway to provide covering clouds so that the team could slither across. Five grenades effectively smoked the area, and Viale led his two men across the span. The rest of the platoon would move across on his orders, or on the orders of the squad leader in case Viale got hit. The three soldiers found a partially demolished wall for protection on the far side and flopped down behind it. When the smoke cleared, in about four minutes, Viale pointed out the three enemy emplacements to his assistants; the first was one hundred yards in their front, and he ordered the BAR man to pour automatic fire into the aperture. The second pillbox was seventy-five yards to his left, and he instructed the squad leader to keep potting away at it with his M1, since it was the least likely to cause trouble to Viale, as he personally went after pillbox number three, twenty-five yards to his right front.

Dragging his carbine in his left hand and filling the inside of his shirt with smoke and fragmentation grenades, Viale crept and crawled into the teeth of the Jap machine gun. It chattered a couple of times as he got within fifteen yards of it. The Nips had spotted his tail bobbing along the ground and had managed to crease the center seam with a low round. Rolling slowly to his side, Viale whipped out a smoke grenade, pulled the pin, tossed it at the gun, whipped out another and tossed it to his own right. In one minute, at the height of the smoke screen, with the Japs firing blindly at him, he crawled sharply to his right, then to his left, avoiding the line of fire. Once out of this fire line he leaped to his feet and ran undetected to the top of the emplacement, which, unfortunately for our antagonists, had no roof. At one-yard range, Viale pumped five shots into three squealing Nips, and checking off the first pillbox of three.

He squirmed back to his two soldiers, told the BAR man to keep ramming bullets at the enemy 100 yards away, and then laid out a plan whereby he and the squad leader would envelop the Japs to their left. Utilizing fire, movement, and the last smoke grenade, Viale moved a few yards forward, covered by his partner's M1, and then the squad leader leapfrogged him, under protective fire of Viale's carbine. The fire was neutralized, and as Viale, from ten yards, poured one full clip of twenty rounds into the aperture, the squad leader got to the rear of the Japs and tossed in two fragmentation grenades, then one more for good measure. They counted the pieces together. Four dead Nips and two Nambus machine guns out of action.

The last pillbox was tough, well dug in with a concrete roof, small apertures, good, clear fields of fire—it defied frontal attack. Viale sent back after his rocket launcher, who came tearing across the bridge as Viale, the squad leader, and the BAR man all fired toward this last obstacle. The bazooka man got into position, fired four times, got three direct hits, and shattered the will of the enemy to resist. It also shattered the enemy, as six more Nips, in various stages of decomposition, were counted around the two machine guns.

The rest of the platoon sprinted across the causeway without incident and immediately went into wedge formation with Viale at the vertex,

Chapter 15: The Damn Fool

moving with his lead scout. He led them up Calle Nueve Street against the nuisances of snipers, the gripping fear of an ambush, and now something new: the plunking in of 90-millimeter mortar shells, exploding with disquieting regularity. A few of his men were nicked, but as they moved toward the source of the firing, the shells went over their heads, and they realized they were under the Nip margin of safety, the umbrella, which was fine for them, but the rest of the company and battalion must be catching hell 500 yards back.

The breeze kept fanning the fires and now Viale had the enemy and the river to his front and left and the conflagration to his right and rear. One escape route remained, east along Dasmariñas Street, and he received word from the company commander that the fire had forced a company change in orders and he was to get the hell out of the "trap." He started moving along Dasmariñas Street and about two blocks away, came to a large junction of Dasmariñas, Nueve, and San Vincente, a junction which he had to cross. Two Jap pillboxes sweeping the junction decided he must not, and these two pillboxes, dug in at the corner of two buildings, defied assault from the street. Viale recognized the suicidal nature of a frontal attack, and began combing the surrounding buildings for vantage points from which to knock out these emplacements. The guns had to be eliminated—and now—if the platoon didn't want a premature cremation. The fires had begun to singe them, forcing them toward the lanes well swept by the two Japanese 30-millimeter automatic weapons.

Viale waved his men to follow him. He circled the intersection and headed toward a house which overlooked the emplacements. Approaching this house from the left rear of the Jap pillboxes, he entered the ground floor. He noticed a rickety ladder leading up to a small window from which, he hoped, someone could throw hand grenades on or into the enemy strong points. Since he was left-handed and the enemy was to the right front of the window, he was a natural to try out his pitching arm. He pulled out the pin on a fragmentation grenade, holding the grenade in his right hand with the safety handle depressed by the right palm. He slowly ascended the ladder and reached the opening. His estimate was correct. From here, the Japs were dead lotus blossoms. As he lifted his right hand to transfer the grenade to his left, a Jap rifleman from God knows where winged him in the upper part of the right arm. The grenade was knocked out of his hand and fell to the ground, now armed, as the handle popped off. In five seconds it would explode. Viale and the men around the ladder all yelled "Duck!" or "Hit the ground!"

since there was nothing to do except fall in any kind of depression, hold your breath, pray, and wait for the fragments to whizz by you (you hoped).

Bob Viale, always the damn fool, jumped to the ground and picked up the grenade, looking desperately around him for a place to throw it. His men were everywhere, in the doorway, in the room, outside of the room. He couldn't aim for the small window because chances were 10 to 1 he'd miss and a lot more men would be hurt when the grenade would explode as it bounced back off the wall. Two seconds—one second—get rid of it, Bob.

Bob never did get rid of it. He jumped to the farthest corner of the room, shoved the grenade deliberately into his stomach, bent double, and blew his guts all to hell. No one else was hurt, which would have been of great consolation to Bob if we could have told him. All we could do was watch as he lay on the floor, writhing and groaning "damn fool" for thirty seconds until he died.

Chapter 16
The Baguio Campaign

FOLLOWING THE MANILA CAMPAIGN, the regiment reorganized and re-equipped in Manila and, on 7 April 1945, received orders to move to Naguilian to join in the attack on Baguio, the Philippine summer capital and one of the remaining Japanese strong points on Luzon. The Regimental Command Post opened at the base camp at Naguilian Air Strip on 9 April 1945. On 13 April, the 3rd Battalion passed through elements of the 129th Infantry near Monglo Hill, on Highway 9 leading to Baguio.

For the next eighteen days, the regiment fought a bitterly contested battle against an extremely stubborn Japanese force. The mountainous character of the terrain was such that the advance was road-bound and offered the Japanese superior positions from which to fight a costly delaying action. From caves in the hills, both on and overlooking the main supply road, the enemy was in position to cause heavy casualties. At the same time, they remained safe from all but direct hits from the air or from artillery.

Once again it was necessary to engage in furious close-in battles to drive the enemy from their strong points. On 13 April 1945, the 3rd Battalion, advancing along Highway 9 from Monglo Hill toward Baguio, employed medium tanks against enemy positions in caves along the road, and reached the west slope of Hairpin Hill after an advance of 1,500 yards against moderate resistance. Mines, encountered along the road during the day's advance, were removed by the engineers. On the following day, Companies I and K attained the high ground at Hairpin Hill where they established defensive positions for the night. During the hours of darkness, several enemy attempts at infiltration were repulsed, with severe losses to the Japanese attackers. Continuing the drive, Companies I and K drove across the high ground and secured Hairpin Hill overlooking the main supply road.

Since the rough and mountainous terrain confined the action to the road, the plan of attack was necessarily limited to a column of battalions. On 15 April, the 1st Battalion passed through the 3rd and continued to push along the road, wiping out scattered resistance.

As the inexorable advance of the Regiment continued, the enemy withdrew, abandoning large amounts of ammunition and supplies along the road. No large-scale enemy defensive positions were encountered until our forces reached the bridge over the Irisan River. Here reconnaissance patrols located what appeared to be a strong defensive line with particularly strong defenses on both sides of the Irisan River Bridge.

On 17 April, the 2nd Battalion passed through the 1st, with G Company leading the attack. By 1035, leading elements of G Company had reached the last turn in the road, 400 yards from the Irisan River bridge. Two medium tanks were supporting the advance. The highway bridge over the river had been blown by the enemy, and the only crossing over the river was by means of a hastily-constructed bypass. As the leading elements prepared to round the corner to assault the bridge, two enemy tanks, one light and one medium, raced around this last corner without warning. On the back of each tank was a platform carrying six enemy riflemen, firing at any target that presented itself. The speed of the enemy attack was so great that the leading enemy tank was abreast of our advance tank before our fire could be brought to bear on it. This tank had mines lashed to it, indicating the intention of its occupants to destroy it and any vehicle with which the tank might come in contact. Pointblank fire from our second tank literally blew the enemy's tank apart, leaving only the treads on the road. The light tank was also destroyed by our tank fire. During this "mechanized banzai attack," enemy fire caused moderate casualties.

The regimental commander, Col. White, who was accompanying the leading elements in the assault upon this key defensive position, was seriously wounded in the thigh. The reason White was so far forward was that the division commanding general had radioed him asking what the hell was taking his regiment so long to get to Baguio. White had then phoned the head battalion, which contacted the lead company, which sent a message to the point platoon, and the answer to the general's query came back in one word: "Japs."

At this slow rate, it would take us another two days to reach Baguio, and the pressure was on from MacArthur's command to finish the mission fast.

Chapter 16: The Baguio Campaign

So Col. White walked over to my dugout at regimental headquarters and said, "Stan, come on, we're going to drive up and see what's keeping our men pinned down so long."

We hopped in a jeep and he drove. He was angry because he suspected that the troops were dogging it. As we drove along the road, we noted the men lolling about; the column had stopped, no firing was heard, and everyone seemed satisfied that they could read some letters, finger their K rations, and catch up on some sleep, fueling the colonel's ire. A few minutes later when we reached the lead company, his fury had reddened his face to the color of his hair.

Spotting the company commander, he hopped out of the jeep, with me about ten steps behind. "What the hell are you loafers doing? Let's get moving, even if you have a kick a few butts to do it."

The captain, who had been through a number of campaigns in the jungle where ten yards advance was a victory, shrugged his shoulders and began to walk slowly, very slowly, toward the lead platoon, which had huddled on this side of the sharp turn, unwilling to creep and crawl around the bend because they had the intuition that mortal danger would be creeping along toward them. Furious, White dashed up to the three riflemen at the point, the little GIs who were supported by thousands of infantry, artillery, airplanes, tanks, quartermaster, ordnance, ships, factories, trucks all behind them and all pushing them inexorably into the muzzle blast of a Japanese machine gun. This was as far as the United States of America extended in this war, and this was where the heroes were, and from here the dead and wounded men would be carried back to aid stations, hospitals, or cemeteries.

White ran up to the trio and said, "Get your asses going around that turn."

"Colonel, sir, there are Japs around that turn. We're waiting for some flame throwers before making the turn."

"Horseshit," said White, "there's not a Jap for a mile," and like all good leaders should, he started round the bend by himself, with the company commander a few steps behind and with me an even more discreet ten feet back.

The moment he made the turn he was confronted by the two "banzai" Japanese tanks with turret guns blazing, and men shooting rifles in our direction. White was hit in the upper thigh by a blast, knocked off the side of the road, landing about ten yards below in a small grassy expanse only a few feet wide. The captain had his head blown off. I dropped my carbine and instinctively jumped down off the road toward what could have been a two-mile drop. But, fortunately, I landed on this same God-given expanse which had saved first White and now me from a death leap. White was in semi-shock and his thigh was gushing blood. I was in complete shock and only came out of it when he groaned, "Tourniquet." Miraculously, and I do not use that word lightly, along with us was the regimental Catholic chaplain, Elmer Heindl, one of the most saintly men I have ever met. While I lay there paralyzed with fear and ignorance, he proceeded to bind up the colonel's leg, probably saving his life.

The firing above us was furious and screaming and groans mixed in with the tank and rifle fire. After what seemed like an hour, but actually was only five minutes, the firing *put-putted* to a stop, the moans rose in volume, and the thank-God voices of American GI's were heard"

"He's dead."

"Finish this one off."

"Get the litter bearers."

"Put some grenades in the tanks."

"Where's the colonel?"

To the latter question I gasped, "We're all down here," and within a few minutes the infantry heroes and medics had put White on a stretcher (for one embarrassing moment I almost lay down in it myself), and then they hand-pulled me up in time for me to witness the colonel being placed on a jeep ambulance, to note the carnage, the dead and dying Japs and Americans, the smoking tanks. We could now proceed to the next bend and maybe some more of the same.

I walked over to Col. White who, though still in shock, and in pain, could whisper, "I guess I was wrong…there *were* a few lousy Japs." Then he passed out, came to again as the jeep was slowly winding down the road, and he waved wanly at me and mouthed what I have since learned is the West Pointer's "goodbye": "It was good soldiering with you."

I didn't return the compliment. I was able to see him again in a Manila hospital two weeks later where he refused to let the doctors amputate his leg. We were next together a year later, after the war, in Washington, with his wife and two children, where he had been discharged from the service, limping badly but standing with military bearing. He had joined our intelligence service which later became the CIA; he was much later named its executive director.

Let's return to the attack on Baguio: during the remainder of the day, the 2nd Battalion pushed forward against heavy resistance from the ridge on the east bank of the Irisan River. Strong Japanese positions on the

Chapter 16: The Baguio Campaign

high ground east of the river made further advances impossible. On the following day, the 3rd Battalion launched an attack, crossed the river, and secured the high ground south of Highway 9, east of the river. From this position, they could bring fire to bear upon the hill mass from which the enemy were holding up the advance of the 2nd Battalion.

Two days of mopping up and of short advances followed, with the 1st Battalion preparing to pass through the 2nd in the assault upon the hill overlooking the Irisan River bridge. With the 2nd Battalion covering, the 1st Battalion launched its attack upon the strongly defended key hill south of the Irisan River, on the left of the highway. After a strong attack, Companies A and C reached the top of the hill and secured the position. During the day and night following, four separate enemy counterattacks were repulsed and the enemy driven off. In one attempted counterattack against Company A's positions, approximately fifty Japanese were observed moving toward the point occupied by the Company Command Post. Only ten men from Company Headquarters were present, and a quick call for reinforcements brought two squads of the 1st Platoon to assist in repelling the attack.

Even though outnumbered, the two squads and company headquarters personnel placed themselves at the crest of the hill, and awaited the company's approach. As the Japanese came within thirty yards of their position, the small force of Americans jumped to their feet and rushed down the slope. Firing as they ran forward, they killed many of the enemy. As the grenade-throwing Americans came within range for close-in fighting, the Japanese, stunned by the ferocity of the attack, broke and ran. Continuing to fire, the Americans drove the enemy completely off the hill, and later counted thirty-five dead Japanese on the slope of the hill. In C Company's area during the same period, three separate enemy counterattacks were repulsed with only minor casualties to our forces.

In four days of the fiercest type of fighting, against a stubborn and well-emplaced enemy, the main Japanese defense line west of Baguio was broken, and the attack was tremendously accelerated. During the four days fighting around Irisan, 430 enemy were killed with only moderate losses to ourselves and thousands of rounds of enemy artillery and mortar ammunition were captured. In addition, three 15-centimeter guns, two 75-millimeter guns and numerous machine guns, rifles, and other pieces of equipment were taken. In spite of a determined enemy and almost perpendicular hills and rainy weather, the regiment had broken the back of the enemy defenses and insured the success of the drive on Baguio.

On 21 April, the 148th Infantry left the road and moved almost due east toward the Trinidad Rice Bowl, to cut the Japanese escape route from the Baguio area. Rapid progress against isolated resistance enabled the regiment to reach the hills west of the Bowl, where patrol action and minor engagements continued until 29 April. Then, in the final action to cut off the Japanese retreat from Baguio north through Trinidad, the regiment crossed the Bowl and secured the hill mass north of Trinidad. Using medium tanks, M-7 self-propelled howitzers, M-8s and M-12s, artillery and mortar fire against stiff resistance, our forces secured the hill mass overlooking the Japanese route of withdrawal to the north. On 1 May 1945, the 129th Infantry relieved the 148th Infantry in that sector.

In this abbreviated campaign, the 148th Infantry killed 1,121 enemy and captured a sizable quantity of enemy arms and material. Among the enemy ordnance captured were one 77-millimeter gun, six 75-millimeter guns, six 81-millimeter mortars, seventeen 50-millimeter mortars, three 15-centimeter guns, one 47-millimeter gun, five 37-millimeter guns, fifteen heavy machine guns, seven light machine guns, and two 20-millimeter antiaircraft guns. In addition, hundreds of cases of ammunition, dynamite, food, and miscellaneous supplies were taken as the regiment drove the Japanese from their prepared positions along the road to Baguio.

Considering the character of the terrain and the subsequent advantage accruing to the defender, the casualty figures for the regiment of five officers and sixty-seven enlisted men killed in action and fourteen officers and two hundred and forty-five enlisted men wounded in action were considered "moderate," unless you knew the eighty-one dead...or were family.

Chapter 16: The Baguio Campaign

Baguio had fallen and the engineers erected an appropriate sign
on the mountain highway

Stanley A. Frankel

Chapter 17
The End of the War

We received news of the dropping of the A-bomb on Hiroshima via a Division weekly newsletter which was mimeographed and distributed to the troops a day after the event. The A-bomb was a surprise to all of us, and, I confess, a pleasant one. It was only in later years that I realized the full horror of atomic weapons. The dropping of the second bomb on Nagasaki was an equally happy read although we didn't quite understand the implications. We figured if two were needed, then perhaps more would be dropped every week. We still expected the Japanese to continue the war until we had invaded and captured their islands.

At the time of the atomic bombings, our regiment was fighting in the Cagayan Valley, driving the Japanese from the main roads and pursuing them in reinforced patrols as they retreated into the higher, more jungle-like areas. They did not have much fight left; their food and ammunition depleted, they only reacted to our patrols-in-force, rarely attempting counterattacks.

On the morning of the day the Japanese surrendered, but before our regiment heard the news, I accompanied a reinforced patrol, commanded by Capt. Herman Lutz. We were motorized, with weapons, carriers, and jeeps. About 100 men in the patrol were sweeping through the side roads from the main artery. As we sped by a small clearing, we noticed a Japanese tank which was battle scarred and apparently abandoned.

Lutz jumped out of the lead jeep, got on top of the tank, opened up the tank turret, and yelled: "There are Japanese in here."

At that moment, the Japanese inside the tanks opened up with rifles and killed him. Within a minute, our troops had surrounded the tank, pulled back Lutz' body lying next to it, and one of our men had dropped a grenade down the open turret. End of story, but it was obviously a tragic one with Lutz having been killed, we learned later, a few hours after the Japanese surrender was broadcast.

We finished the patrol, encountering no more stragglers, and I went directly to Division Headquarters to make my report to our regimental commander, Col. Schultz, who was temporarily camping near those headquarters. Schultz wasn't around when I went into his tent to report, so I walked to the latrine, about 200 yards from his quarters, and in the dark, dropped my pants and sat on one of the two-seaters. The other was occupied by what I thought was another American officer. I began small talk and his response in the dark sounded strange, sort of "Japanese-y." And it was! As my eyes grew accustomed to the darkness, I made out a Japanese officer sitting quietly next to me. He was weaving his head up and down, actually respectfully bowing.

I got the hell out of the latrine, ran into Col. Schultz, and screamed: "There's a Japanese in there."

"I know," replied Schultz. "The Japanese surrendered this morning, and several top Japanese officers came into Division Headquarters to make the surrender formal. You have just met their colonel."

Soon, I also met two other Japanese officers, one the general, who commanded the Japanese troops in the valley. They were docile, smiling, subservient. They said their men had been ordered to stack arms, and were waiting for orders to move to whatever prisoner-of-war camp we would provide. They also stated phlegmatically that hundreds of their soldiers were starving and sick and that many of these could not walk. We offered to provide trucks for the sick and wounded and starving, and I won't ever forget their cool response:

"Don't provide trucks. Those who can't walk will be left there to die. Not to worry!"

At this cold-blooded offer, our own general blew his top and ordered that every last Japanese soldier would be brought out, the weakest first. The Japanese officers quickly agreed, helped organize the truck convoys, and a ferrying of prisoners began, with the sick, wounded, and starving being loaded first.

I wrote an account of one of these truckloads being brought to our camp, and present it here, as written, in the heat of the surrender with memories of Japanese atrocities to our sick, wounded, and starving still fresh, and festering. I entitled this story "The Bloody Remnants."

TWO 6X6 ARMY TRUCKS, crowded with seventy-eight Japanese prisoners, bumped along Highway 5, headed for Cabagan, Cagayan Valley, Luzon, where the 148th Infantry Regiment was maintaining its prisoner of war stockade. Several kilometers from the town, the Filipinos

Chapter 17: The End of the War

spied the trucks and started screaming with glee. The screaming followed the trucks along the highway into the city. Slowing down to five miles per hour, the vehicles were tailed by hundreds of men, women, and children insulting the Nip prisoners in Tagalog, Ilocan dialect, and pidgin Japanese. Someone started throwing rocks at the Nipponese and soon the trucks were running a gauntlet of flying stones. Unfortunately the two guards and the drivers of the 6x6s as well as the Nips were struck by the rocks. Capt. Griffiths in the front seat cab of the lead truck ordered the drivers to speed through the town. When the trucks arrived at the prison camp which he commanded, he reported the incident to the regimental commander. In strict accordance with the Geneva Convention as set down in the Rules for Land Warfare, the regimental commander directed that, in the future, the army guards would take any and all steps necessary to insure the safety of themselves and their prisoners. Under this directive, rampaging civilians might get something more lethal than stones thrown back at them.

These malarial, lice-ridden, underfed soldiers were the stinking remnants of the Imperial Japanese Army in Luzon, commanded by infamous Gen. Yamashita. They were now limping into our outposts from the jungles and mountains, then trucked to our prisoner of war stockades. Among the mounting hundreds of these bowlegged, scrawny sons of heaven were a few who had participated in the Death March some years back. There were many others positively identified as having committed brutalities not sanctioned by the Rules of Land Warfare. These yellow people were the same breed of Japanese warriors who had bayoneted fifteen wounded men of this Regiment in an ambush on Zanana Trail, New Georgia, in July, 1943; who had castrated two of our noncoms on an outpost at Hill 700, Bougainville, Solomon Islands in March 1944, before our costly counterattack restored our lines; who had lined up 136 Filipino men, women, and children at Tondo District, Paco Lumber Yard, Manila, in February 1945, tied their hands behind their backs, and then successively raped, machine gunned, bayoneted, and burned them exactly eight hours before our regiment liberated that district.

There weren't many combat infantrymen remaining in our regiment who had begun the bloody trek to Manila two years back on New Georgia, but both old timers and the newer replacements nursed grudges of long and short standing. By and large, the men were not entirely in accord with the policy of humane treatment, but realized in their disagreement that the Rules of Land Warfare would be rigidly followed regardless of their personal feelings.

One staff sergeant squad leader, whose entire twelve-man squad was wiped out in hand-to-hand, room-to-room fighting in the Legislative Building in Manila in February favored turning the stockade over to the Filipinos. "Maybe we're too kindhearted to give these bastards what they deserve but the Flips won't screw around with them very long."

A replacement corporal who had joined us for the mountain fighting at Baguio and Balete Pass, April of this year, first wanted to kick a few hundred of them in what we shall euphemistically call the lower groin.

His lieutenant, who won the Silver Star for blowing four Nip machine gunners to Shinto Heaven with a well-placed grenade a few months back, wasn't quite so hard boiled: "You can't kick a guy when he's down. What they should do is put about two squads of veteran infantrymen along with a few hundred Nips and let them go to it with bare fists. That's one detail I'd volunteer for."

One grizzled battalion commander, a follower of the Halsey school of philosophy, kept the discussion on a higher plane. His thesis: "Kill the bastards, anyway you want, but kill them." His battalion had stormed the Japanese mountain positions around the Irisan Bridge at Baguio, 17 April, and he had won that high ground, making Baguio's fall inevitable, all at heavy cost to his Battalion.

A major, the regimental S-2, had a more cynical, attitude which began to prevail over crass emotionalism: "The Japanese enlisted man is a dirty little animal, naive, uneducated, highly disciplined. He will do exactly as his officers wish, without exception, or Buddha help him. When he rapes and plunders and tortures it is because those actions are condoned and encouraged by his officers. The Japanese officers, from Yamashita on down, are entirely responsible for the brutalities their men have committed. Those officers should be tried and hanged. The men should be sent back to their rice fields."

There were a few apologists for the Japanese, mainly among new officers who had never heard a shot fired in anger; still, their impersonal reasoning couldn't be laughed off. They imagined that some of our own soldiers were no angels, and that under the pressure of smoke and bullets and fatigue they too had committed atrocities. True, out of vindictiveness or bloodlust, some of our soldiers had cut loose. Wounded Japanese were given the coup de grace by litter bearers who got tired of lugging them to battalion aid stations under sniper fire. Dead Japanese, still warm, had lost their gold teeth to GIs with a good dropkick technique. Looting by

Chapter 17: The End of the War

our soldiers and by the Filipino civilians in Manila often made Japanese plunder tactics seem small potatoes. Nowhere did the enemy ever acquire the trucks or the finesse to empty private homes down to the last electric light fixture.

We returned much of the loot when the bullets stopped zinging. Higher headquarters inspectors decided it was now safe enough to check combat infantrymen with silk underwear dangling from combat packs and Bakelite ashtrays rattling around in mess gears. However, the thousands of bottles of Canadian Club and scotch were beyond extraction, and the typewriters, refrigerators, radios, and golf clubs had mysteriously ended up in sacrosanct headquarters sections or officers' clubs. Several of us picked up a 1941 Buick Super Deluxe in the garage of Puppet President Jose P. Laurel's home, to save it from the looters. Realizing that such pretentiousness was not consistent with our lowly ranks, we presented it to our regimental commander with appropriate ceremony. We assumed that his West Point background and eagle insignia would afford it ample protection. After the fighting had subsided, our gift was rudely jerked away by holier-than-thou Corps MPs who stated that civilian cars could be retained for military use only. Two weeks after that, we noted the Buick with official sticker being driven by a WAC sergeant for an engineer general. We were all a bit puzzled about the military use to which a base engineer general could put the car to which a combat regimental commander couldn't.

Getting away from looting, let's return to the treatment of the Japanese POWs.

GI visitors to the stockade invariable softened. At first they glare at the sickly internees who grinned happily and bowed unctuously at their captors in appreciation for the vegetable hash, the rusty stove, the moldy squad tents, and the nearby creek where they are permitted—nay, ordered—to clean up. This Japanese GI was a model of docility and gentleness, and he took his anti-dysentery bismuth and his anti-skin ulcer iodine with gratitude and humility. One who spoke good English told us that all of them would have surrendered long ago if they had known of this kind treatment. They claimed their officers kept them fighting with tales of American brutality, and many of them had reserved one hand grenade at all times for honorable suicide in fear of the aftermath of capture. One endearing charm was the way these yellow fellows took their Atabrine tablets, pills which ere so vile tasting that our men even washed them down with GI lemonade. These Japanese put the three tablets on the tip of their tongue, sucked the yellow things with apparent relish, and let the pills dissolve in their mouths. Then they bowed their thanks.

Most of our soldiers went away shaking their heads in the firm conviction that these puny individuals are children who have been badgered and browbeaten by fanatical officers. A short look-see at the officers' section reinforced this belief. The officers, in accordance with the Rules of Land Warfare, had additional privileges: special quarters, Japanese orderlies, cots, and better-prepared food. They still retained some of their arrogance, and they surrendered with the understanding that they were officers and expected to be treated like officers. The liaison captain from their Brigade Headquarters explained this carefully to our regimental S-2 prior to surrender. Upon entering the stockade, one lieutenant wanted to know where the hot showers were. Another lieutenant, the only one that was ever solicitous as to the welfare of his men, requested that Filipino native labor be recruited to dig their latrines and garbage pits and to put up their tents. He claimed his men were too weak, which they might have been. However, they did their own internal construction, in accordance with those rules.

The enlisted men were completely cowed by their officers, who completely ignored them except to turn half-heartedly the snappy, stiff salute, which is always tendered by the soldiers, regardless of their physical condition. When one Japanese captain entered our hospital ward, the eleven sad sackomotos in various stages of decomposition jumped or tried to jump to attention. Our own aid man ordered them to lie down when it appeared that the Japanese officer would just as leave have them stand that way while he had his arm bandaged.

In addition to the notes about the Japanese prisoners, we ran into other unrelated incidents which, to a limited extent, tied us into the other war which had ended in Europe some months back. These notes were written outside the largest city in the Cagayan Valley where we had mopped up at war's end:

> "A batch of German refugees passed by yesterday ...had been bombed out...the second time they have lost their homes in ten years. In 1935, most of them fled here from Germany...most of them are Jewish ...and I was able to give them food and transportation back to the rear where they can get additional food and shelter.
>
> Also had another experience with another German group who arrived several hours after the German Jews. Thinking them

Chapter 17: The End of the War

also German Jewish refugees I began to chat with them...and then discovered they were actually Nazis...who had been in the German Diplomatic service, assigned to the Japanese in the Philippines and living the life of guests under Japanese rule. Of course, we were required to treat them like decent people...even tho [sic] they all maintained that Germany was not whipped yet. I still haven't in me the bitterness to kick a person when he is down...as these Germans now were. But I did lead them out of sight before some of the more excitable Jewish soldiers could get at them. When their leaders thanked me profusely for all I had done for them, I felt guilty...as if I had betrayed my coreligionists.

The Japanese in the Philippines surrender
to the 37th Division commanding general and staff

Stanley A. Frankel

Chapter 18
An Unexpected Assignment

At war's end, I had been with the 37th Division since before war's beginning, and, combined with my combat experience, had enough points to have earned immediate rotation home. Most of my comrades who had shared my experiences were already in or on their way back to the States. Obviously I was looking forward eagerly to setting sail for San Francisco.

It was not to be. As I have noted, I had never wanted or expected to be a combat infantryman, and even after being commissioned I had regularly requested a different assignment only to be summarily turned down. After the month-long New Georgia campaign, I typed out a Request for Transfer to *Yank Magazine* and sent it through channels to the commanding general of the 37th. In a few days the Request bounced back: "Transfer denied. There is a drastic shortage of experienced infantry officers."

Again, I knew why, which was the reason I wanted out. Even though along with the rejection came promotion and medals, I was not consoled. Up the Solomon ladder we went, next to Bougainville, followed by another Request for Transfer, a request denied for the same reasons, then Manila, request submitted and denied...a mountain fight for Baguio...more consolation prizes, but no transfer.

Now, with the war over, I wanted only to return to civilian life. Instead of orders to return to the States, however, I received a summons from commanding Gen. Robert S. Beightler.

In his Division Headquarters tent, he returned my salute with a warm handshake and a generous comment: "Maj. Frankel, I have been aware of your interest in writing, and I felt bad having to turn down your many requests to be transferred to a writing assignment. However, you know there was a shortage of qualified infantry officers. Now I am happy to tell you that I have decided to honor your request."

"But general...the war is over...I didn't make any recent request. I want to go home."

Stanley A. Frankel

"Major, someone has to stay here with my staff and me for the next few months to assemble and draft the history of our division, and we have chosen you. After all, major, you realize that there is a peace to be won.*"*

What winning the peace had to do with me I shall never know. All I wanted to write now was "Goodbye Army," but my latest fate delayed that message by another three months.

I was ordered to remain in the Philippines along with a small group of editors and writers to work on the history of the 37th Division. We returned to the States in early 1946, received our formal discharges in March 1946, and continued working on the history evenings and weekends, while we adjusted to full-time civilian jobs. By the end of that year, our assignment had been fulfilled. The original draft then began its rounds to the commanding general, various division, regimental, and battalion staff officers, and to company commanders. Facts were checked and re-checked, debated, and disputes finally resolved. The 37th Infantry Division in World War II *was published in 1948 by the Infantry Journal Press in Washington, D.C. It consisted of 400 double column pages of words, maps, and pictures.*

The response was as good as I could ever have imagined—generally accolades. There was a large sale of the hard covers, and eventually a soft-cover reprint. There were very few beefs about errors (of which there have to be many) and not a single complaint about mistreatment from any soldier from the corps commander to the lowly rifleman.

My foreword is reproduced below:

THIS HISTORY was written by 40,000 men. Half of the writers were from Ohio, and the second half came from forty-seven other states. Some wrote their paragraphs in a bold, heroic hand, and some wrote simply and effectively. Thirteen hundred of the writers were killed as they placed periods at the end of the most brilliant chapters.

The rank of these historians ranged from buck private to major general; their ages from an unofficial 16 to a concealed 55; their civilian vocations from janitor to banker; their God from Christ to Buddha; their ancestral homes from Edinburgh to Tokyo.

These 40,000 men have a good story. It is not unlike stories by other men who pushed a common enemy around in other parts of the globe. It concerns a gallant band of American soldiers who trained and fought and bled, and, in some cases, died. Those who trained and fought and bled, and the kin of those who died, hold in their hearts and minds everything worth preserving. For all of them, and for their families and friends and for those to come, we are trying to preserve their story in printed form.

Chapter 18: An Unexpected Assignment

We editors didn't write this history; we merely reproduced it on paper. The reproduction falls short. But it is as faithful and as complete as our space and our skill permit. Many paragraphs remained unreproduced; they remain untapped in the hearts and minds of our soldiers. Other inspiring chapters lie buried with their writers in Guadalcanal, in New Georgia, in Bougainville, and in Luzon.

We promise our 40,000 writers that this reproduction of their story is the very best we can do.

Many soldiers helped magnificently in getting this history ready for the printers. Our assistant editor, Tech. Sgt. John K. Macdonald, unselfishly volunteered to remain in the army seven months after his normal discharge date to work on the job. A large part of the reproduction is his.

When Sgt. Macdonald was discharged, Chaplain Frederick Kirker, longtime division chaplain, took over. He tied together the loose ends, rewrote sections which last-minute information had rendered incomplete, and put the final history into the hands of the printers. His contribution to the 37th Division in its written history—and in its actual history—was tremendous.

Maj. Gen. Robert S. Beightler spent hours in proofreading, checking, and advising. In peace as in war, he demonstrated how close the 37th Division is to his heart. On Luzon, Brig. Gen. Leo M. Kreber, then chairman of the Division Historical Board, assisted the editors materially, as did Col. Russell Ramsey, who was appointed the new board chairman when the Division was deactivated. This Historical Board was chosen from all division units. It met periodically to settle literary disputes and transact business and consisted of the following men: Lt. Col. Chester Wolfe, Lt, Col. Sylvester Del Corso, Maj. Grank Middleberg, Maj. John C. Guenther, Master Sgt. Richard W. Gohn, Master Sgt. Walter J. Freeman, Master Sgt. Paul Dearth, Staff Sgt. Herbert C. French, Tech. 4 Joseph C. Caserta.

While overseas, the historical staff was large and invaluable. On Luzon, the following writers, reporters, research men, typists, and artists contributed efficiently to our task: Tech. Sgt. Jack F. Ehlinger and Tech. Sgt. Harry F. Storin, Jr., both of whom voluntarily remained in the army for three months after returning to the United States to work on the history; 1st Lt. David Andrews, whose skilled writing embellishes the New Georgia and the Bougainville chapters; Capt. Francis X. Shannon, chief artillery writer-adviser; Maj. John Guenther, whose outline of the Division history from induction through Bougainville was extremely helpful and whose intimate information on division staff procedure was generously shared with the editors; Capt. Charles H. Holzinger, specialist on the planning phase of the Philippine operation and the surrender period in the Gagayan Valley; Sgt. Jesse Gauthorn, who did a large portion of the art work for the history; 1st Lt. Shelby Minton; Tech. Sgt. Kenneth E. Chronister; Tech. Sgt. Virgil R. Bates; Staff Sgt. Harold L. Geisse; Staff Sgt. Perry Tenenbaum; Sgt. Bernard R. Flesher; Tech. 4 Harry B. Bunting; Tech. 4 Eugene J. Ruane; Tech. 4 Eric Adler; Tech. 5 Keith L. Barker; and Tech. 5 Harry Lipowitch.

Our appreciation to these assistants and to unnamed others is keen.

And to those 40,000 writers who permitted us to reproduce their story, we are eternally grateful."

Chapter 19
Return of the Heroes

In our return to the United States, I ran into some rather unexpected reactions to the war...soldiers...the uniform. As I re-read my comments, which I'll reproduce here, I understand the bitterness felt by the Viet Nam veterans who came back from a horrendous, bloody, unwinnable war to be greeted with derision, scorn, and unconcealed distaste. By comparison, our reception was relatively congenial. But, at the time, it was hard to digest. Here, then, is a description of the way it was:

"WHEN THE DEVIL are you going to take those damn clothes off?" an old friend greeted me. "I feel a revulsion every time I see a uniform."

I had stepped off the gangplank of the USS Veltervreden on 4 December after 3½ years of stiff combat overseas. On the 5th I bought some officer's woolen clothes, the first I had ever worn since I was commissioned overseas. With pride I tried on the coat, pinned on my ribbons (the first I had ever put on), and polished my major's insignia.

On the 6th, I walked in on this old friend and was greeted, in all seriousness, by a man who, I have since found out, was merely symptomatic of the revulsion, the indifference, the apathy, and the damn short memories of the American public.

As mentioned above, my ship docked at 7:00 p.m., 4 December, at San Pedro, California. This ship carried the remnant of the 148th Infantry Regiment of the 37th Division, the most decorated regiment of the most decorated Army Division in the Pacific. In the 148 was Tech. Sgt. Cleto Rodriguez, who killed eighty Japanese with a BAR in Manila four months ago. There were Bly and Brown; Marok and Gall; Downey and Josephs; and hundreds of others who, during the past forty-two months, had stoically accepted heroes roles a dozen times from New Georgia to Bougainville to Manila to Baguio to Balete Pass.

Men like these walked down the gangplank, greeted by two civilian photographers, one white-haired, flag-waving spectator, and a small GI band bored with the proceedings. The war had been over for three months, and the public no longer gave a damn. Each day a few thousand men docked at these ports, and the law of diminishing utility had immediately affected the market for heroes. The royal welcome was kingly only in its complete and exclusive isolation.

The man who had a revulsion for uniforms was replicated each hour of each day. As the busloads full of homecoming GIs rolled toward camp, one face smiled and one hand waved on the thirty-mile route. She was a prostitute, undoubtedly, because when the bus sped on by, she quickly closed her mouth and dropped her hand. Poor business prospects.

The cross-country trip was a revelation. At one stop, the men jumped out to snatch some sandwiches. As one of the soldiers fumbled in his watch pocket for change, the counterman snapped, "War's over, soldier. You're not on the gravy train any longer. Better come up with the cash." This gravy train GI wore the Purple Heart with two clusters and the Silver Star.

America was fed up with its heroes, and its heroes sensed this apathy and hastened into civilian clothes, where possible. The USOs were closing down or slowing down. The famed canteens were shutting up shop. The newspapermen didn't bother with Joe anymore, even if the quiet Joe who came home late had fought four times as hard and as long as headline Joe who came home early. All of which was all right. But what hurt was public indifference: the smile, the handshake, the clap on the back, the well-meaning questions, and interest, and the look of admiration. America had once shown these things to her heroes. But her sense of timing was bad. She had been kind and heroic and helpful while the war was on. She had heaped her spiritual wealth and her cheers on the soldiers who were home, soldiers who came in two categories: (1) men back from overseas who were ashamed of cheers because their greatest gift, they knew, was in being home while their buddies were still fighting, and (2) men training to fight who hadn't heard a shot fired in anger and who should have been even more ashamed at the free drinks and the tinsel.

Now, the war over, the buddies of category 1 were coming back, entitled to their share of the nation's gratitude, and the men who were only training were coming back after having earned that acclaim, only to learn the irony of America's sense of timing. America in her hustle and

Chapter 19: Return of the Heroes

strikes and short memories didn't know the hurt she was doing. Her welcome was bitter and inadequate when most deserved.

One of my corporals, wearing a DSC, met a cop who was writing him out a ticket for parking too long in one spot.

"Have a heart, officer," my man said. "I just got back from overseas. Give me a break."

The cop had heard this before: "Too damn many crybaby heroes around this place for me. You know the war's over don't you? If you don't, this'll learn ya."

One of our most heroic platoon leaders, a 1st Lt. from Bronxville, New York, who had received a battlefield commission for leadership, wrote me this:

> I bought myself the spiffiest looking officer's uniform I could find…just one since I had about 100 days terminal leave and then would have to put it away for good. I dolled up in all of my medals, and then called on my gal. She threw her arms around me and then stepped back and said: "You poor kid. Let's go down right now and buy you something decent to wear." Decent, I said? Listen, Mary, I feel awfully decent in the uniform of a U.S. Army officer, and right now I'd feel indecent in anything else. Well, we went round and round, and it ended when she started to bawl and handed me my ring back and told me, "You can come back with that ring as soon as you get rid of your war nerves." I give up.

I don't think these latecomers resent the inadequacy of the present-day veteran's job-finding organizations or the red tape of the GI Bill. The men realize that these aids are something new, that they are growing, and that the veteran must be patient. What he does resent is this atmosphere of business-as-usual vacuum which pervades the restaurants, the streets, the bars, the basketball games, the shops, places which must have expended so much good will on ill-deserving or reluctant soldiers during the war that now, while the returnee is both deserving and willing, they have none of the stuff left.

I saw a staff sergeant limp up to a ticket window in Columbus, Ohio, the other night. He put down a quarter for a ticket and was told off by a gum chewing ticket salesgirl: "We cut that service rate out a month ago, buddy. Where have you been?"

The sergeant reached into his pocket for a second quarter and casually apologized: "I'm sorry. I've been at Camp Pendleton being fitted for a wooden leg."

Then it might work in reverse. My technical sergeant, assisting me on my present job of drafting the division history, told me this one:

> He called up an exclusive nightclub at the edge of town to reserve a table for four. He reserved it in his name: Technical Sergeant Jack Ehlinger. The voice at the other end commented: "You're lucky the war's over. Before that we never accepted reservations from enlisted men, only officers. Now there ain't quite as many so we'll hold a place for you."

I had lunch with one of my platoon sergeants the other day. Overseas I censored his mail, and got an insight into his romantic life. He told me that his wife-to-be wrote him at the end of each letter: "All I want is you. If you get back, I'll never want anything anymore." He got back, and they were immediately married. In our conversation, he laughingly told me that he throws that quotation up to his wife whenever they squabble, and she, no damn fool, has a pat reply: "You know I didn't mean that. I only wrote it to keep up your morale so you would feel like fighting and protecting us back here."

A soldier now getting off a transport after one, two, or three years in the foxholes and pillboxes must have that feeling about America. The cheers and radio bravadoes and gracious letters from acquaintances and the chummy notes from the boss…were these things merely a means to an end, merely a hypocritical, cowardly gesture to keep him fit mentally for the contortions ahead? Now that the end was accomplished, did America in her heart still know and appreciate his sacrifice? Or had America washed her hands, set up professional veteran aid groups as a sop to the soldier's political strength and dismissed Johnny Doughboy as a good whose necessity has gone, and therefore whose sacrifices can be paid for out of taxes instead of out of the heart?

Another civilian sitting at a bar offered a cigarette to his friend sitting next to him. The friend took the smoke, and jerked his head in the direction of the soldier sitting on the other side, as if to say: "Why don't you offer him one, too?"

The civilian jammed the package in his pocket and said in a stage whisper: "Why the hell should I give him anything, I've got a brother in the service."

Chapter 19: Return of the Heroes

That attitude, too, is common. People with brothers, sisters, fathers, sons, daughters, mothers, and even nephews in the service agreed that the serviceman should be well treated, but should be well treated by George. Let George do it, because George hasn't got a relative in the service. He doesn't suffer because a loved one is in danger, so George should go all out for helpfulness. The sad part is that there are damn few Georges, and everyone has someone close who is just out of the service, but that fact does not confer the inalienable right to sublimate yourself in your brother's GI shoes and thus slough off the courtesies which you believe are shown to your brother (and you).

Maybe my little niece summed it all up. I proudly pinned on 7-year-old Ann a combat inf. badge, my proudest war medal.

She looked at it for a few seconds, then tried to tear it off her sweater.

"What's the matter, Ann? You used to like my medals?"

Uninhibited, Ann shot back: "I don't want yer ole medals, anymore. Why dontcha get me a wristwatch like Billy's uncle got him?"

Another GI with a chestful [sic] of ribbons, a combat infantry badge, and four overseas stripes, left a 10¢ tip for a waiter in a medium-priced New York restaurant.

The waiter muttered to me, half in anger and half in warning, that I better not be so stingy: "The lousy bastard. Just because he'd done a little free sightseein', he thinks we enjoy doin' things for him, for nothin'. When are those lousy heroes goin' to wise up?"

That GI, by the way, wore the insignia of the 32nd Infantry Division, and he had seen the entrancing jungles of New Guinea and the sights of the treacherous Villa Verde Trail in Luzon.

At the Golden Gloves the other night, a black corporal from the 93rd Division, which had fought on Bougainville and New Guinea, stood rigidly at attention while the Star Spangled Banner was playing.

"Look at the nigger showin' off," the civilian next to me snapped out in the middle of "through the perilous fight." Then, at "the land of the free," he went on: "Give a nigger or a Jew a uniform and right away they begin to strut worse than Storm Troopers." This democrat, of course, merely poured a little of his vitriol of intolerance on his revulsion for uniforms.

Maybe now is the time to sponsor "Be Kind to Soldier Week." There are still millions in uniform and millions to come back from overseas. I don't think the soldier is going to ask for a thousand dollar bonus or a new home or a free meal or even a beer on the house. Not yet, that is, although he might be driven to get his material reward what he failed to get in plain, ordinary friendliness. What he would like to have is:

"Your Division had a brilliant record."

"The air corps (or the navy or the infantry or whatever force the guy represents) made one of the greatest contributions to the war effort."

Or "That Silver Star you're wearing didn't come cheap. That much I know."

A man in the streetcar wryly read the caption to a picture of pickets, noting that fifteen ex-servicemen, in their discharge uniforms, were walking with placards. Probably most of them wore their uniforms for the effect; a few might not have been able to get "decent" wool clothes.

"Look at the sons of bitches," groused the man in the streetcar. "Usin' that sympathy gag. As if it would do them any good. Anybody with sense nowadays knows that no one cares a rap about a soldier suit. Those things went out of style in September."

A sergeant working in the reenlistment section, Fort Hayes, told me this: "It's a funny thing, but lots of the soldiers who join up ask me if they can be sent someplace away from their home towns. I think they are ashamed of being in uniform where their friends can see them."

An officer in this same section explained to me that the enlistment offers every possible advantage for a young man: security, stimulating work, travel, good living. But the major obstacle to many reenlistments, one of which can't be reduced by the enlistment board, is the prevailing animosity of civilians toward anyone in a serviceman's uniform. "It's as if the civilian has always been conscience stricken on seeing soldiers. During the war, many of them masked this in gushing hospitality and generosity. Now, with no war pushing them, they react just the opposite, as if the man were walking around in a uniform just to aggravate them. Craziest thing I've ever seen."

Chapter 20
Afterthoughts

FRIENDS AND FAMILY, probably having run out of small talk, have occasionally asked me how my war experiences affected me. Behind their question is the implied query: how can a peaceful, somewhat cowardly, fairly genial and affectionate man, fresh out of valedictorian graduation at Northwestern, and giver (as well as taker) of the Oxford (pacifist) oath, have come through five bloody battles in one piece, mentally and physically? Hasn't shooting and being shot at, sleeping for weeks in a muddy foxhole, going to the bathroom in an open field, eating C-rations for weeks on end (mainly meat and vegetable hash, occasionally heated for variety), hasn't all of this changed your values, your character, your personality?

Yes and no. I recall a song from the post-war musical comedy "Call Me Mister" titled "Still a Jerk." The lyrics describe a young jerk—stupid, insensitive, boring, crude, unlikeable—who's drafted into the infantry and goes off to war. For three years, he endures the hell of war, the shot and shell, the bravery of comrades, his own courage under fire, and the final victory (as the music crescendos patriotically). And, goes the song, he emerges from this traumatic, intense, heart-and-back-breaking experience "Still a Jerk."

Yes, those war years changed some things, just as the passage of any half-decade does something to an individual. I hate picnics. I refused being a Boy Scout leader when my kids joined the Scouts. I detest Spam. I cannot accept waste, including wasted food, having observed starving civilians in Manila lined up along our garbage cans to eat what we threw away. I used to believe there was no such thing as a good war or a bad peace. I now believe WWII was a "good" war.

And no, I am still pacifistic, and as far as I'm concerned, WWII was our last "good war." I think Korea was a marginal affair and Viet Nam was obscene, and I led Businessmen for Peace in Viet Nam very early in that horrible fight and earned being named to Nixon's Enemies' list for my upfront, outspoken opposition to Viet Nam.

I have retained my sense of humor, dark humor, perhaps, and as hard as I tried, neither then nor now could I despise the Japanese people or soldiers; their leadership, perhaps, but their soldiers were brave and, though outnumbered, out-supplied, and outgunned, they never quit. Went about doing the job they were ordered to do, just like my men and me.

I recognized that Machiavelli was correct—the ends do justify the means—and I understand the terrible atrocities committed by both sides because, in a war, like in an alley fight, you must win, and you do everything you think will aid the victory, regardless of morals, ethics, honor. The Japanese treatment of our captured soldiers and of the civilians they conquered was, in their eyes, necessary to win. Our bombing of non-military targets climaxed by the use of the A-bomb, twice, was done in an effort to speed up the victory, shorten the war, and, in the end, or so we told ourselves and believed, save the lives which would have been lost if the war dragged on.

I cannot bring myself to feel as objective about the Germans. Their leaders, their soldiers, their people. I do not believe the Holocaust was conceived and implemented as a means to win their war. Quite the contrary—the diversion of their soldiers, transportation, and construction materials needed at Auschwitz and Belsen and elsewhere actually hurt their war effort. The ironic loss to the Germans of scientists like Einstein, Meitner, and Rabi kept them from winning the race for the A-bomb. No...these cruelties had no Machiavellian justification. They were not necessary. They didn't do what they had to do to win; they did what they didn't have to do, and therein lies my inability to forgive Germany...ever.

As I wrote in the introduction to this book, I did resort to one trick to alter, just a bit, my own perspective and values. During the war I kept a diary (against army regulations). This diary was in my back pocket throughout the fighting and each night before trying to sleep, I'd make a few entries about the day's activities. I managed to hide these diaries and brought them home with me in the bottom of my duffle bag. As the years went by, say on June 15, 1953, I came home after a rough day at the office, fighting my boss, yelling at my fellow workers, irritated and aggravated by some business setback. Then I would go directly to my diaries and look up July 15, 1943, exactly ten years back. I would read the entry which often went:

Chapter 20: Afterthoughts

> Today, my best squad leader, John Pierce, was killed by a sniper; and by the time we returned from patrol, it was too late to heat the hash; and my foxhole had to be dug hastily into coral, which made it shallow and hard to sleep on; and tomorrow at dawn we attack entrenched Japanese machine guns on our right flack and we'll lose more men...and maybe, even me.

Reading that entry, all of the petty, trivial problems I had at the office no longer bothered me. From this new perspective, I understood what was really important, and what was, as the GIs would say, "chicken shit." Then, at dinner, when I recounted to Irene the unpleasant events at the office, she'd shake her head and ask why I seemed so happy and unruffled. This device has remained my secret, until now, for I confess that, on those few occasions when Irene and I would squabble over doing the dishes or clearing the table or cleaning the drawers or disciplining the children, I would have those magic diaries to smooth my feathers and remind me that life is too short to get high blood pressure from the insignificant irritants of life at the office or at home.

Do I have bad dreams? Sort of. But my recurring bad dream has to do with going into an exam in college and suddenly realizing I forgot to study for the exam. My worst and most recurring army dream? Not the shot and shell and blood and guts. No...I do still have a dream that I have been called back into service, and, as I begin to sweat over how the hell I can avoid returning to the military, I wake up, and two Tums return me to a more pleasant, dreamless sleep.

All of the above is the effect on me as a person. How about what the Army did to me professionally?

It is true that at age 25 I was a major—a field grade officer "managing" hundreds of men. Since the war I have been a boss—a manager—of varying numbers of men and women, but since I have tried to be a behavioral type manager, given to participatory management, consulting with my underlings, listening to their ideas and permitting them, at times, to talk me out of my original position, I find my army leadership lessons ones I had to unlearn. The army is an example of classical management, where the higher one's rank, the more important one is, and when an order is given, it must be carried out, without rhyme or reason. The soldier's instructions are "to do or die; not to reason why."

I must say I understand why orders have to be carried out in the army, without hesitation or equivocation or even thinking about those orders. But that's not the way to run a corporation, and the two management styles are at opposite ends of the management-theory pole.

The other things the army taught me: close order drill; digging foxholes deep and fast; disassembling and then reassembling and then firing light machine guns and mortars and Browning Automatic rifles; marching stiff and straight "hut-two-three-four"; taking an Atabrine pill every morning; dousing a mess kit in boiling water; carrying toilet paper in my left rear pocket; awakening to a bugle call at dawn; saluting and returning salutes; sleeping under a pup tent or in a bomb shelter; and censoring mail. All of these skills, forever after, are useless. Thank God. Often, during the war, as I lay in my tent or in my foxhole, I pledged that, if I ever got out of my military predicament in one piece and with relative sanity, I would never again complain, and I would try to erase all the bad memories.

Maybe this book is the last erasure. Re-reading the letters to Irene, I came across the following paragraphs written in the middle of the war, which opened up once again some of my thoughts on fighting in general, for whatever this position might be worth:

Dear Irene—

Please forgive the rambling tenor of this letter…just let me get it out of my system: In my mind war is the worst of all atrocities. There is no such thing as a moral war or a just war or a war with rules. War is for keeps and any means whatsoever which can be used to gain victory must necessarily be employed. Thus, I don't condemn Japanese atrocities from a moral standpoint. We who are bombing German women and children (incidental to winning the war) have no right to preach ethics of war.

My own condemnation is of the sense which inspires these atrocities. Thus, if such brutality as maiming a man would help the Japanese win the war, then they would be justified in their crimes. So, if for one minute I thought that bayoneting Japanese children, raping their women, and cutting their starving soldiers in half would help us end this war and save American lives, I'd say bravo and let's get to it. I believe bombing of cities will break the German will to resist. That is a mass atrocity but justified because it will help end the war. I believe that individual atrocities such as rape, murder, starvation of individuals will not

Chapter 20: Afterthoughts

help the war effort, but instead will steel the determination of our enemies. So, I say, without one thought to the ethics of the situation, blow off their legs with 1000 pound bombs but do not drive a bayonet through a child's intestines.

There is no such thing as a moral code in this all-out war. We must win and we must commit every act which will help us win. Actually the Japanese are damn fools for their crimes. The treatment of the prisoners of Corregidor have made us angry and that treatment has also made many of us want to die fighting rather than surrender. Naturally, we have felt that way all along. We take few prisoners; the Japanese take none.

Our own brutality does not make us angels. Many of our boys kill defenseless Japanese…starving Japanese…diseased Japanese …Japanese that they could very easily take prisoner. They kill them under the guise that the Japanese might be feigning sickness (which has been true in some cases.) However, I do not condemn this treatment of prisoners as bad from a moral standpoint. I condemn this brutality because a live Japanese prisoner can give us information which might save American lives. The condemnation, therefore, is strictly one of ignorance rather than of brutality.

Summing it all up, I guess I'm saying that in a war Machiavelli was right. The end justifies the means. There's only one logical, acceptable end when you are fighting a war. That's to win. Whatever it takes. Better not to quibble about who is more brutal, more inhuman; better to figure out how to put an end to wars. Once war is on, it's every man, woman, and child for himself and herself; and let the devil take the hindmost.

Following up the abstract philosophy in my letter to Irene, may I make a few more personal and subjective comments:

The ultimate question all soldiers must ask themselves is how they managed to survive when fellow soldiers all around them were being wounded or killed. That question I put emphatically to myself innumerable times. How had I not only survived but never even been scratched when men to the right of me and the left of me and to the front of me and to the back of me were shot up?

Maybe I was more scared than most, or more prudent, or more adept at digging foxholes faster and deeper. Or maybe my analytical skills were so honed that I could anticipate dangerous situations and avoid them.

Maybe somebody up there was looking out for me (which I do not believe for one moment).

More likely the answer was luck. Blind, dumb luck. How else can I account for what I think of as perhaps my nearest miss? One night early in the battle for Manila the Japanese were shelling our positions, firing at will in the knowledge that we were not firing back with our artillery nor hitting them with our bombers. Gen. MacArthur had ordered no counter-battery fire, since so many Filipinos and American prisoners were trapped along with the Japanese troops that our indiscriminate firing would kill as many friends as foes.

Five of us were huddled around a table in the living room of a small house a few miles from the Pasig River studying a map to pinpoint our crossing of the river on a small barge the next day. Explosions had ringed our house every quarter hour, but nothing had yet been close enough to cause casualties. Suddenly a shell—probably a 60 mm.—came through a shrouded window and hit directly onto our table. We were all knocked back by the force of the explosion.

I thought I was dead, but as I recovered consciousness, I began to feel parts of my body to make certain I was intact. My legs, my arms, my face, my nose, my back, my stomach, my groin. I remember crying out in excited surprise: "Hey…I haven't been hit!" I did not know that I was calling out to three dead men and to a fourth dying in the hallway. His feet were kicking spasmodically when I ran over to him and saw his intestines splattered all over the floor. He died a few minutes later.

The only other survivor was a Filipino civilian, owner of the house, who had been in the next room and had caught a small shrapnel fragment in the stomach—just a flesh wound.

I think that all of us had learned the lesson early in combat that we should never make close friends. Amid all the camaraderie, between-battle poker and volleyball, amid all the laughing and crying and insulting and comforting, always live at arms length—in a cocoon—with a protective shell between you and everyone else.

Yes, you feel badly at the loss of your "friends," but—and I'm a bit ashamed to confess it—your first reaction when the guy next to you is hit and you are unscathed, is: "Thank God, it's not me."

The heroes whom I've cited in this book had the opportunity to act and were capable of transcending selfish instincts by exposing themselves to the shell or the bullet that would have killed or wounded their comrades. That kind of Medal of Honor–instinct suggests, if nothing else, a kind of divinity. Maybe in each of us there is a tiny bit of Rodger

Chapter 20: Afterthoughts

Young or Bob Viale. I don't think I have it, but I thank God (or anyone else) that I was never put to the test…and that my blind luck prevailed.

Stanley A. Frankel

Appendix

One dictionary synonym for appendix is addendum—another, supplement. A secondary definition in the Random House Dictionary *goes: "The short tube at the bottom of a balloon bag by which the intake and release of buoyant gas is controlled."*

Leaning heavily toward the secondary definition, the appendix of this book contains quite a bit of material, all relating to the author's life and writing, but only indirectly or not at all related to the subject of this book.

Please forgive the uncontrolled release of buoyant gas, but the author suggests that your skimming through the appendix might produce some laughter or tears or general interest, a kind of, forgive the mixed metaphor, frosting on the cake.

Stanley A. Frankel

Awards

Regtl. Commander L. K. White passing out medals after the Battle for Manila, assisted, to his left, by the author

Transcript of Letter to Dr. Marc Hollender

Dr. Marc Hollender, my ex-roommate and longtime friend, kept all the letters I wrote him during WWII. Recently, he sent me a package of fifty-three of them, and this one is, I think, one of the more amusing.

Stanley A. Frankel

27 May, 1945
Philippines

Dear Marc,

Your book, *People On Our Side*, just arrived, and I am afraid it is going to interfere with the war effort...at least my own contribution which, between us, doesn't make a helluva lot of difference anyway.

For the past six months we have been driving so hard that I just got out of the book habit. I carried my pocket edition of *Time* and *New Yorker* in my pocket, hastily read a story in between air raids, chow, or early morning plumbing, and carried on. In fact, now that we have become a bit more settled, I just haven't had the patience to stay with any one book. However, this one by Snow really sounded good, and I started in on it during a lull in my paper work...and had one hard time putting it down. Thanks ever so much, both for your thoughtfulness and for your good taste.

I have received letters from you while on the move, but we are not p[e]rmitted to hang onto these letters...security...in case we are found dead with them...so I burn yours, memorize the address...and promptly forget it...necessitating my writing through Irene. This time, I think I have that address licked...and if I send this one through Irene, you'll know the address was lost during the last five minutes before typing it on the envelope.

We have engaged in very strenuous operations, and even through I am not digging out the Nips with my bayonet, I have had my moments...at which I would have sold out my chances for survival for 10% pre-war prices. While ducking a truckload of Nip artillery, mortar, and rocket shells at Malacañan Palace in Feb., I tried to figure out a way to get into the Aid Station...the one really safe place in the Palace... surrounded by double walls...wand in a bomb proof shelter. So, seeing a couple of wounded men lying around the area, I figured that they were my ticket into the aid station. Thought I: ...carry one with you and they'll have to let you in. Once in, let them try to get you out...even if you have to stretch out with the corpses over on the right. So, selfishly, I made the 100 yards to the wounded in four seconds, looked around for the lightest guy and found a semi-midget, picked him up, got to the aid station, was practically dragged out again by a three man litter team who needed a fourth...so I volunteered...with one guy on each of my arms and the third with a gun in my back...and thus I made a couple of more trips...which caused me extreme nervousness and exhaustion...and the

Appendix

upshot of it all is that my Regtl. Commander, watching from afar, so he couldn't read my expression, put me up for a Silver Star. Doubt if I'll get it because the Awards Board at higher Hq. knows that I am not the hero type and something is fishy…but it just shows how ironic a [illegible] a guy's instinct for self-preservation results in.

And if you think I am fooling in the last paragraph you are dead wrong. I did not jump; I was pushed.

Anyway, maybe this new discharge plan will affect me someway. I have 98 points, will make 103 if I can get this award; that's a lot most places, but in my regiment with men who have been wounded five times (25 points); with just as much time oversees, in the army, and number of campaigns…plus a DSC or a SS tossed in…well…I rank only in the upper third of the officers; and I don't visualize "Santa Claus" Marshall breaking up this team just because we happen to have the required number of points. And Marc, this is a real team…with five good fights under our belts, way over 100% casualties and almost 100% medals for heroism; several companies cited by the President, and the Regiment up for one for Luzon; that's a tough combination. The Regiment has killed 18,657 Japs these last three years, and they really smell bad. Which is about my major contribution…since my platoon leading days I am no longer an eager beaver.

Nothing else, Marc, hope this war-mind of mine hasn't bored you…but it's tough to live something for three years without it becoming a part of you…for better or worse.

Thanks again,

Stan Frankel

A Presidential Distinguished Unit Citation is awarded to those units (be they squads, platoons, companies, battalions, regiments, divisions) who performed in battle, as a unit, which, if an individual soldier so performed, would have resulted in the soldier receiving the Distinguished Service Cross.

Following is the War Department citation of the 148th Infantry Regiment, for its performance in the Philippines, Jan. 9, 1945 until March 4, 1945, particularly for the regiment's key role in the capture of Manila. The author was an officer in the 148th Infantry Regiment during that period.

Stanley A. Frankel

Distinguished Unit Citations

The *148th Infantry Regiment* is cited for outstanding performance of duty in action against the enemy at Luzon, Philippine Islands, from 9 January to 4 March 1945. In every phase of the campaign in which it participated, the *148th Infantry Regiment* achieved spectacular success, carrying out its missions with courage and speedy efficiency. By its capture of the critical road junction of Plaridel by spearheading the drive into Manila from the north, advancing 137 miles in 24 days, by its liberation of the Americans interned at Bilibid Prison and the patients and refugees at the Philippine General Hospital, by establishing, under fire, the vital bridgehead across the Pasig River and by the major role it played in destroying the fanatical Japanese garrison in Manila, the *148th Infantry Regiment* contributed immeasurably to the brilliant success achieved by the United States forces in the Luzon campaign. In every engagement, the regiment exhibited outstanding combat efficiency by uniformly inflicting severe losses on the enemy, while sustaining only moderate casualties. Each unit of the regiment performed its assigned duty with consummate skill and fidelity. Service troops worked unceasingly to supply the combat troops, the medical detachment performed innumerable acts of gallantry in caring for both soldiers and civilians, and the cannon company gave invaluable direct fire support. Over open ground, through city streets, the *148 Infantry Regiment* met and decisively defeated the enemy wherever he chose to make a stand. Its brilliant combat record is a tribute to the courage and skill of every man in the regiment and exemplifies the finest traditions of the military service. [General *Orders No. 34, War Department, 10 April 1946.*]

Also receiving the Presidential Distinguished Unit Citation was the author's first infantry assignment, Co. F, 148th Infantry Regiment, for its heroism in Bougainville. Although the author was not a member of Co. F during this action, many of his first platoon and fellow officers were a part of this heroism, and a number were killed or wounded in the action.

Company F, 148th Infantry, is cited for the magnificent gallantry, heroism, teamwork and will to win that it demonstrated in this crucial operation, and for its tremendously significant part in the action on Hill 700, which resulted in a victory of major importance to the entire United States defense of Bougainville Island.

Appendix

During the recent offensive action by Japanese forces against the United States positions on Bougainville Island, *Company F, 148th Infantry Regiment,* participated in a counterattack against enemy positions atop Hill 700, which resulted in the destruction of Japanese forces in that sector and the removal of a major threat to our positions.

This action, which took place on 12 March 1944, was a double envelopment by Company E and *Company F, 148th Infantry Regiment,* and represented one of the outstanding examples of daring and courage to occur in this theater. The proximity of the enemy to our lines prohibited the use of supporting artillery, and the rugged terrain precluded the use of tanks. Our attacking forces were compelled to advance against almost every conceivable obstacle. The enemy enjoyed a commanding position, excellent fields of fire, superior observation, and the natural advantage accruing to the defender. *Company F*, on the other hand, had to execute a difficult flanking movement, across precipitous, fire-swept terrain, against a determined and confident enemy occupying strong defensive positions.

The attack, begun simultaneously with the advance of Company E on the east flank, was a charge against enemy positions under a withering hail of fire at point-blank range. Utilizing rocket launchers, flame throwers, smoke grenades, and other infantry weapons, the men of *Company F* swept over the Japanese positions, made contact with Company E approaching from the east, and secured the objective. *Company F* lost three enlisted men killed, and four officers and 39 enlisted men wounded in this assault. The enemy lost 407 counted dead in this immediate area, and were practically annihilated. The backbone of the entire enemy offensive on Bougainville was broken. [General *Orders No. 50, War Department, 17 June 1944.*]

SNAFU Revisited

Speaking of Awards, forty-five years later, the author's tribulations in trying to collect some of his medals led to a recounting of the experience in a column he wrote for This Week Magazine. *Here's the story:*

SNAFU...Situation Normal All Fouled Up. It's been 45 years since millions of us left the WWII armed forces...and that acronym. But, events of the past months have reassured me that SNAFU is alive and well...perhaps with a dash of Murphy's Law sprinkled in.

About a year ago, my 37th Infantry Division Monthly Newsletter advised WWII veterans of the Fighting 37th that if they had not received any medals, earned in those bloody 3½ years in the S. Pacific, they could write to Army Personnel, St. Louis, Mo. and get them. All of us, at various occasions these past 45 years, had worn our ribbons, the colorful, paper-clip-size symbols of our medals. But the metal Medals

...the beautiful silver and bronze, heavy and designed hanging decorations

...had not been issued. So, because I have an 8 year old grandson in his Army-GI Joe stage, passionate for live medals, I wrote to Army Personnel, identified myself as having served 5 1/2 years in the Armed Forces in WWII, the last 3 1/2 years overseas and in combat.

Two weeks later, a shocker-letter arrived. Personnel advised me politely but firmly that a careful checking of their files did not turn up any record of my ever having served in the Armed Forces, in WWII...or any War. When I reported this to a friend...who suddenly became an ex-friend...he snidely opined that my name was probably filed under "Coward"...and no medals due.

In truth, I became the Unknown Soldier...not known for ever having fired a shot in anger in any battle fought by my country. Out of my enraged subconscious drifted my Army Serial Number...which I had not thought of for 45 years: O-1794932. Quickly, I dashed off a note to Army Personnel giving them this dogtag identification. Two weeks later, a shoebox-size container was delivered by the mailman, full of dozens of plastic wrapped and some beautifully boxed medals. Many I knew and recognized...the Bronze Stars, Oak Leaf Clusters, Combat Infantry Badge, Presidential Citation, Sharpshooters Medal, Good Conduct Award...and on. In addition, there were many I had never heard of...Philippino [sic] Victory Medal, Asiatic Campaign Medal, Solomon Islands Freedom Medal, Conqueror of Manila Award, Liberation of Luzon...and so forth. The less important the medal, of course, the more dazzling.

The following weekend, we were invited to that ex-friend's house for dinner, and for the first, and last, time in my life, I wore all the medals. The combined weight didn't help my hernia, but revenge was sweeter than the pain they caused. Then, I put them back in their containers and delivered them to my grandson, who was ecstatic. He pinned a medal onto the pajamas, and wore several to Trinity School. His teacher was impressed, not enough to permit him to wear them all day, but enough to ask Adam to recite what his Grandfather had done to win each one.

Adam, having been subjected to my somewhat embellished and not-so-modest tales of bravery, almost every weekend the past two years, did himself and his Grandpa proud.

End of story? EOS? Not on your life. SNAFU? Of course. For, every two months thereafter, an identical box arrived, conveying the identical medals…again and again and again. I could have gone into the second-hand medal business. Instead, I used these duplicates and triplicates, once at my nephew's wedding, highlighting the pre-nuptial dinner with a speech-making ceremony, pinning a combat infantry badge on the blouse of the bride and the Good Conduct Medal (billed as "an advance") on the groom. Others to the children in the wedding party. Just two weeks ago, another box arrived from Army Personnel

…this one, to my disappointment, was small…the size of a deck of cards. I unwrapped the little package to find it contained only one medal. A letdown. Except, this one, as I read the inscription, was for…hold onto your seatbelt…. "*Bravery in KOREA.*" I had never been within 1000 miles of Korea; was not in any way involved in the Korean War.

Now…I realized that my name must have been cranked into the computer, and I live day to day, expecting my medal for bravery in Viet Nam … Granada … Panama. Or…perhaps they'll jack the whole process up a notch or two and toss me some really significant Awards

…like the DSC…Distinguished Service Cross…or…as the end of SNAFU nears…maybe even that long coveted Congressional Medal of Honor.

Then…and only then…will I write to Army Personnel, asking them to shut off the spigot. My guess is that their reply will read something like this: "Dear Mr. Frankel: We have searched our files and we cannot find any records indicating that you have ever served in the Armed Forces of the United States in any war."

As long as they don't ask for their medals back!

Medals are usually earned, though the author cautions the general public that often woven into the formal citation is that germ of truth surrounded by some degree of hyperbole and literary license. This is particularly true, says our author, about awards he received, mostly a "poetic" extension of the truth. In any event, with those disclaimers, here are some of the awards recommended for our author and/or actually awarded to him. He urges those readers who find immodesty disturbing to skip over the next three or four pages.

Stanley A. Frankel

Headquarters 148th infantry
Office of the Regimental Commander
APO 37

15 June 1945

Subject: Award of the Legion of Merit.
To: Commanding General, 37th Infantry Division, APO 37.

1. Under provisions of AR 600-45, it is recommended that the Legion of Merit be awarded to Captain Stanley A. Frankel, ASN 0-1794932, Regimental Adjutant, 148th Infantry Regiment, for exceptionally meritorious conduct in the performance of outstanding services at New Georgia, Guadalcanal, and Bougainville, Solomon Islands; and Luzon, Philippine Islands from 18 July to 3 March 1945.

2. DESCRIPTION OF THE ACT:

a. Date—18 July 1943 to 3 March 1945.

b. Place—New Georgia, Guadalcanal, and Bougainville, Solomon Islands; Luzon, Philippine Islands.

c. Narrative—During the campaign at New Georgia, Solomon Islands, Captain Stanley A. Frankel, then First Lieutenant, distinguished himself by his outstanding services as Executive Officer of Headquarters Company, 148th Infantry. Captain Frankel repeatedly displayed fine leadership in leading his unit against well-concealed enemy defenses in the dense jungle interior. By constantly checking and applying all security measures to the utmost advantage, he counteracted the natural advantage that the dense foliage afforded the enemy. Whenever his company bivouacked for the night, Captain Frankel efficiently assisted the Company Commander in arranging the perimeter defense and interior security. On many occasions he increased the efficiency of carrying parties going to the forward elements by personally directing their movements over the muddy trails. Captain Frankel devoted much time to the coordination of the movement of the limited transportation available, in order to assure the front-line companies of an adequate supply of precious water.

On one occasion a wire party led by the Assistant Communications Officer was halted by an enemy ambush consisting of a light machine gun team and several riflemen. Locating the approximate position of the enemy weapon by its distinctive sound, Captain Frankel led his men off the narrow trail to within 30 yards of the enemy machine gun. While he had his men lay down diverting rifle fire, Captain Frankel made his way alone over an alternate route in the direction of the communications party. This movement exposed him to the extreme danger of being shot at by the enemy or by uninformed friendly troops. After successfully making contact with the communications party, Captain Frankel maneuvered it out of its precarious position to safety.

With the successful conclusion of the New Georgia campaign, the 37th Division returned to Guadalcanal. In early November the division resumed military operations against the enemy at Bougainville, Solomon Islands. Captain Frankel remained in command of the 148th Infantry rear echelon consisting of about 100 men. On 16 November 1943, replacements for the 37th Division, numbering 30 officers and about 600 enlisted men, arrived at Guadalcanal. As it was not intended to send these replacements forward for a period of three weeks, Captain Frankel, on his own initiative, submitted a plan to the Commanding Officer, 37th Division rear echelon, for training these replacements during the coming two weeks. Based upon his experience in the New Georgia campaign, Captain Frankel outlined an intensive two-week training schedule covering jungle warfare and including combat firing problems, lectures on jungle tactics, and field exercises. Captain Frankel organized the replacement officers and enlisted men into three temporary companies and supervised the training program, in addition to his regular duties as Personnel Officer. The realistic and specialized nature of this training program fitted these men for jungle warfare, and they were easily absorbed into the division's regiments on Bougainville. Captain Frankel was highly commended for his excellent achievement by a regimental commander.

Near the end of the training period, 26 November 1943, Hell's Point Ammunition Dump was ignited and at 1:30 p.m. shell fragments began to land in the rear echelon area. Captain Frankel decided to evacuate the area as a precautionary measure. Ordering a formation of all replacements and rear echelon personnel, he instructed an officer to march the men along the beach away from the ammunition dump to a clearing 500 yards east of a nearby Naval Construction Battalion and to await further orders there. Captain Frankel then made an inspection of each tent, discovered several soldiers in their tents, and ordered them to join the departed troops. By this time the explosions grew louder and two shells burst almost simultaneously in the area. Captain Frankel, nevertheless, continued his inspection until he was sure that everyone had left the danger zone. As the explosions mounted in intensity, shells began to fall in the new troop-assembly area. The number of military personnel had been greatly augmented by hundreds of men from all units, including natives, marines, Australians, army and navy troops. Captain Frankel succeeded in maintaining effective control among his men despite the prevailing confusion. He immediately issued orders to his officers to lead their men across the Tenaru River 200 yards further down in order to safeguard his men from injury by flying shell fragments. Upon reaching the river bank, Captain Frankel noticed several men floundering in the water and, with the aid of several volunteers, he succeeded in pulling the men to safety. He then had all the men who were unable to swim form a chain by holding one another's hands, and led them over a shallow sand bar where the water was only shoulder deep. By this method about 400 men crossed the river safely. His cool and intelligent leadership was responsible for a swift and orderly evacuation of over 600 men under particularly difficult and hazardous conditions.

In March 1944, when the Army announced the rotation program for men who had been overseas a specified length of time, the Regimental Commander inaugurated a program for training new men to replace key non-commissioned officers throughout the regiment. Captain Frankel requested that he be permitted to train administrative personnel as replacements for those clerks eligible for rotation. Organizing what was then the first formal Clerks' School to be held in the Southwest Pacific, Captain Frankel set up a two-week program of lectures and practical work in army administration for 60 men who possessed qualifications for this type of work. This school proved to be eminently successful, and one year later 43 of the 60 men were engaged in important administrative work throughout the regiment. Captain Frankel was highly commended by higher headquarters on this innovation, and several other units requested and received permission to have some of their enlisted personnel attend the school.

During the entire period that Captain Frankel was Personnel Officer, all his work was performed in a superior manner. After two inspections by higher headquarters, the 148th Infantry Regiment was commended for the up-to-date completeness of the personnel records, commendations which reflected Captain Frankel's meticulous attention to duty and his administrative ability.

Appendix

On 28 January 1945 during the Luzon campaign, Company L, 148th Infantry, was ordered to seize the towns of Mexico and San Fernando. It was imperative that internecine strife be prevented if the regiment's progress were to continue unimpeded. Traveling through 14 miles of territory which had not yet been cleared of Japanese, Captain Frankel caught up with the motorized company and entered San Fernando with the advance elements of the strong combat patrol preceding the column. He immediately contacted the two opposing factions, pointed out to their leaders that any conflict that directly or indirectly endangered the security of American troops would be dealt with summarily, and then outlined a compromise which was readily accepted by both factions. This skillful diplomatic action facilitated the entry of the regiment's troops into the town, and hastened the advance on Manila.

On 5 February 1945, it became necessary to evacuate 1275 civilian and military prisoners from Bilibid Prison in Manila. Fierce fires set off by enemy demolitions were approaching the front of the prison yard and the lives of the prisoners were endangered. In addition, the flames were forcing the Japanese toward Azcarraga Street, immediately south of the prison, and the Second Battalion, 148th Infantry, was engaged in numerous fire-fights around the prison walls. The Regimental Commander assigned Captain Frankel as assistant to the Regimental Executive Officer who was in charge of the evacuation. Captain Frankel helped organize a truck convoy and expedited the leading of litter cases on trucks and ambulances. After making a thorough inspection of the hospital wards, despite the fact that he was subjected to enemy machine gun fire, to assure himself that no one was left behind, he led the convoy five miles to the Ang Bay Shoe Factory. He then returned to assist in supervising the withdrawal of combat elements of the Second Battalion, and departed from Bilibid Prison only after all troops had left the area. His prompt action materially contributed to a swift evacuation of all the internees.

Captain Frankel has been recommended for the Silver Star for gallantry at Malacañan Palace on 7 February 1945, when he rescued two wounded soldiers during an intense mortar and artillery barrage, and for establishing and maintaining a forward command post on the south side of the Pasig River under heavy enemy fire, on the following day. Captain Frankel's keen judgment, extreme devotion to duty, and outstanding leadership have been vital factors in maintaining the highest standards of both combat and administrative efficiency in the 148th Infantry Regiment.

3. The service of Captain Stanley A. Frankel has been honorable since the acts on which this recommendation is based. This recommendation is based upon the attached statements of four eye-witnesses. I have personal knowledge of all facts herein contained not included in supporting certificates.

4. Captain Frankel's home address and next of kin are: Mrs. Olive Frankel, (Mother), 1215 Amhurst Place, Dayton, Ohio.

D. E. SCHULTZ,
Lt Col, 148th Inf,
Commanding.

A TRUE COPY
R. BOLINGER Capt, 148th Inf,
Per. Off.

Appendix

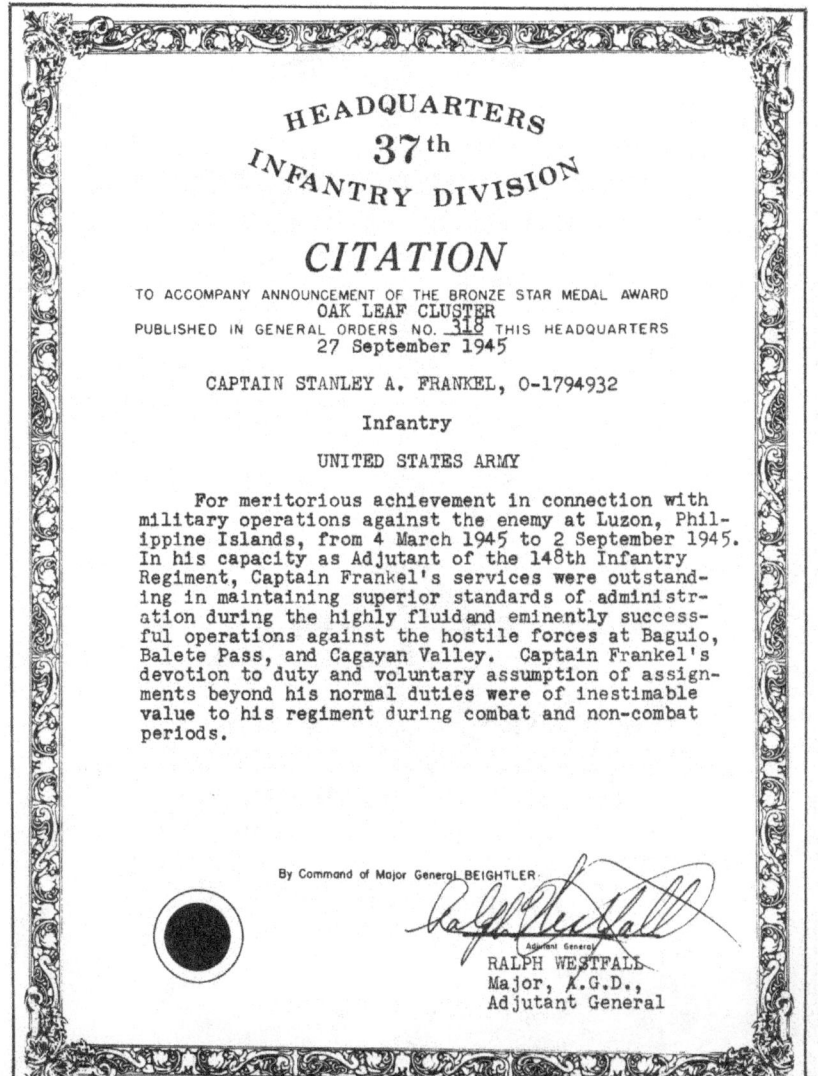

Stanley A. Frankel

THE UNITED STATES OF AMERICA

TO ALL WHO SHALL SEE THESE PRESENTS, GREETING: THIS IS TO CERTIFY THAT THE PRESIDENT OF THE UNITED STATES OF AMERICA AUTHORIZED BY EXECUTIVE ORDER, 24 AUGUST 1962 HAS AWARDED

THE BRONZE STAR MEDAL
W/"V" DEVICE

TO CAPTAIN STANLEY A. FRANKEL, UNITED STATES ARMY

FOR heroic achievement in connection with military operations against the enemy at Luzon, Philippine Islands on 7 and 8 February 1945. Captain Frankel, a regimental adjutant, left his sheltered position during an intense enemy shelling of the regimental command post area in order to rescue two seriously wounded men who were lying in an exposed area. His courageous actions in this instance saved the lives of the two men. Later, while under heavy enemy fire, Captain Frankel crossed a river with two men and a small wire-party to establish a forward Command Post, which enabled the Regimental Commander to maintain uninterrupted control of the tactical situation.

GIVEN UNDER MY HAND IN THE CITY OF WASHINGTON
THIS 26th DAY OF July 19 88

THE ADJUTANT GENERAL   SECRETARY OF THE ARMY

THE UNITED STATES OF AMERICA

TO ALL WHO SHALL SEE THESE PRESENTS, GREETING: THIS IS TO CERTIFY THAT THE PRESIDENT OF THE UNITED STATES OF AMERICA AUTHORIZED BY EXECUTIVE ORDER, 24 AUGUST 1962 HAS AWARDED

THE BRONZE STAR MEDAL

TO MAJOR STANLEY A. FRANKEL, UNITED STATES ARMY

FOR meritorious achievement in ground combat against the armed enemy during World War II in the Asiatic Pacific Theater of Operations.

GIVEN UNDER MY HAND IN THE CITY OF WASHINGTON
THIS 6th DAY OF January 19 89

THE ADJUTANT GENERAL   SECRETARY OF THE ARMY

Appendix

THE UNITED STATES OF AMERICA

TO ALL WHO SHALL SEE THESE PRESENTS, GREETING: THIS IS TO CERTIFY THAT THE PRESIDENT OF THE UNITED STATES OF AMERICA AUTHORIZED BY EXECUTIVE ORDER, 24 AUGUST 1962 HAS AWARDED

THE BRONZE STAR MEDAL
WITH SECOND OAK LEAF CLUSTER

TO   CAPTAIN STANLEY A. FRANKEL, UNITED STATES ARMY

FOR   meritorious achievement in connection with military operations against the enemy at New Georgia, Guadalcanal, and Bongainville, Solomon Islands, and Luzon, Philippine Islands, from 18 July 1943 to 4 March 1945.

GIVEN UNDER MY HAND IN THE CITY OF WASHINGTON
THIS 26th DAY OF July 19 88

THE ADJUTANT GENERAL   SECRETARY OF THE ARMY

This is a rare picture which might well be captioned: "The Sublime and the Ridiculous," for it features, on the left, Maj. Gen. Robert S. Beightler, commanding general of the 37th Infantry Division throughout the entire war, and, on the far right, author Stanley A. Frankel, at that time a buck private. In the middle is writer, John R. Tunis, who visited the 37th Division in training at Camp Shelby, Miss. to do an article for *Look Magazine* on a national guard outfit being prepared for combat, shortly after the war began. Frankel, who knew Tunis, had been chosen as liaison between the 37th Division staff and Tunis.

Birth of a Patriot
*How a smart youngster came to realize
he'd been too smart for his own good*

by John R. Tunis

Following is an article which Tunis did for the N.Y. Herald-Tribune Magazine *about, not the general, but Frankel. Make of it what you will, and don't forget the word: hyperbole.*

It was the spring of 1939. I had just returned from three months in the tragic continent of Europe when I was asked to talk to the student body of a large Midwestern university. So there I was, trying not very successfully to make those thousands of eager-faced boys and girls see what was coming. And I asked them to make up their minds, before the drums started to roll, whether they'd fight. And if so, what for.

Were there any questions? There were. A fellow rose from the crowd, slender, slight, bespectacled. Even before he spoke I guessed he was president of the debating society, and probably an undergraduate leader. He had, he said, only two questions.

First: Was I a veteran of the World War?
Answer: Yes. Second: Was I a member of the American Legion?
Answer: Yes. He sat down, very pleased with himself.
Applause.

In other words, I was an alarmist. I was a warmonger. I was the older generation. I was convicted out of my own mouth. There were no more questions.

Afterward he came down to the platform, one of a group of pressing youngsters below, and announced somewhat truculently that he was a pacifist. It appeared he was not fighting in any "imperialistic war." And so forth.

Later I heard from him by mail. He turned out, as I'd suspected, to be president of the debating society and also president of his class. He was alert, intelligent, articulate. We corresponded at some length after his graduation that June, when he got a job as a "leg man" on a Chicago daily. In the fall he heard of an opening in the publicity department of a broadcasting chain in New York. He wrote me about it. Luckily I knew one of the men in charge, and could recommend him. He got the job. And delivered, as I expected.

All the while, events in Europe were slowly forcing him to change position. Though at the time he hardly realized this. The Russian-German pact in the summer of 1939 shook him badly. The invasion of Norway in the spring of 1940 was a blow. The defeat of France that spring really upset him. He was confused and distressed. Then in the fall of 1940 he was drafted.

He went reluctantly. He was leaving a good job. He didn't enjoy Army life, Army food or Army officers, and his letters every week concealed nothing. That winter of '40-'41 was far from amusing. No equipment. Nothing much for an active mind to do. Life was grim and monotonous.

Last spring, however, the tone of his letters changed a little. Equipment started coming in. He was over his basic-training period. He talked with some slight scorn of the new draftees who expected "swimming pools and hostesses." His division got organized and equipped. Last summer they went on maneuvers in Louisiana. When Stanley left Camp Shelby he was a recruit. When he returned he was a soldier.

He was a soldier but he was still not mentally prepared for the task ahead. Then one sunny afternoon in December it came. Like that. The Japs were on our necks. The next day so were the Germans and Italians. Several weeks later he wrote me. This letter was different.

"I guess you think I've been a damn fool. I wasn't really. I was just too smart for my own good. Back at college we thought we had war all figured out. The big shots started the wars—and stopped them—and

Appendix

made their millions out of them. While the little guys fought in them—and paid for them.

"So we figured we'd toss war out the window. The warmongers wouldn't fool us again with that old gag about making the world safe for democracy.

"Then this war started. For a while it was like all the rest. But gradually it dawned on us that this was different. This time the little guy was fighting his own war. Not to save democracy in some far part of the world, but to save himself from slavery.

"Now we know. All of us. Every guy in this camp and others I know about. We've got to win this war. It's for us—it's for everything that means anything to us. It's for the very things we thought we were fighting for when we were campus intellectuals.

"I wish I could write. I'd like to write a letter to all the older people who've been wondering about the younger generation. They can stop wondering. We're all right. Hitler and his gang better watch out."

Someone ought to write it, Stanley Frankel. As long as you can't, do you mind if I do it for you?

One of my staff sergeants who was tremendously helpful to me at war's end in writing the History of the 37th Division was Jack Mac-Donald, now writing impressive weekly columns of material for columnists and disc jockeys as well as conducting his own radio-TV show and putting together an occasional book of his own. Jack sent me the following "An Old Colonel's Laws of Combat" which, like Murphy's Law, has more serious truth in it than funny fiction. Jack was a combat infantryman with the 148th in WWII and can attest to these truths, the actual source of which we do not know:

A wise old army colonel, with plenty of infantry experience, put together some "Laws of Combat" for new young officers. They may have more validity than anything to come out of the training manuals. The colonel's "Laws" include:

1. Anything you do can get you shot, including doing nothing.
2. The only thing more accurate than incoming enemy fire is incoming friendly fire.
3. Body-count math is three guerrillas plus one probable plus two pigs equals 37 enemy killed in action.
4. Friendly fire isn't.

5. Things that must be together to work are never shipped together.
6. No combat-ready unit has ever passed inspection.
7. If the enemy is in range, so are you.
8. The easy way is always mined.
9. If you are short of everything except enemy, you are in combat.
10. When you have secured an area, do not forget to tell the enemy.
11. All five-second grenade fuses will burn down in three seconds.
12. If you are forward of your position, the artillery will fall short.
13. The enemy diversion you are ignoring is the main attack.
14. If you take more than your fair share of objectives, you will have more than your fair share to take.
15. When both sides are convinced they are about to lose, they are both right.
16. Professional soldiers are predictable, but the world is full of amateurs.
17. Don't look conspicuous—it draws fire.
18. If your attack is going really well, you are in an ambush.
19. If it is stupid but works, it is not stupid.
20. Never share a foxhole with anyone braver than you are.
21. When in doubt, empty the magazine.
22. Never forget your weapon was made by the lowest bidder.

Speaking of colonels, Col. Lawrence K. White, second of my three superior regimental commanders, has shown up in this book from time to time, and his comments on my published account of the famed (bloodless) Battle of Balintawak may interest and amuse you:

As the Regimental Commander of the 148th Infantry who "took" Balintawak Brewery in the Philippines, I was particularly interested in the account written by my staff officer, Major Stanley Frankel, which appeared in your Mar. 9 issue (of This Week Magazine).

May I take the liberty of adding a few more details which may amuse your readers.

On the whole it is pretty accurate. With regard to Gen. Krueger, Commanding General of the 6th corps he actually did show up. I remember well standing on the bank of the river with Krueger and General Griswold as we observed the soldiers wading the river holding their helmets full of beer over their heads. Gen. Krueger asked why they were carrying their helmets that way. I told him they were full of beer. He then said, "What's all this shooting I hear just up the road?" I explained that we had a pretty good fire fight up around the Bonifacio monument. Then he said, "Well, what do you think about soldiers drinking a helmet of beer and going right into a fire fight?" I said "Sir, I don't know but I wouldn't like to try to stop it." Gen. Krueger then said "Well, I don't think I would either."

Having finessed that one pretty well I said I was then going to cross the river and go into Manila on foot. We could not yet get any jeeps across. Gen. Krueger than said to me, "As soon as you get in try to make contact with the First Cavalry. I haven't heard anything from them since they got into Santo Tomas yesterday." This I did by taking a Buick away from a Filipino who had just stolen it. He agreed to drive me to Santo Tomas which he did as I stayed prone in the back seat with my forty-five in his back ready to carry out my promise to take him with me should anything untoward happen enroute.

Stanley A. Frankel

Appendix

Aye or Nay?

In the introduction, the author mentions his pacifism as well as Time Magazine's *recognition of his efforts while in college to keep the U.S. out of the war. Here's a copy of an excerpt from that* Time *article.*

Every leader knows that to fight a war, whether for conquest or in self-defense, he must give the young men of his nation a cause so good and just that they are willing to be ripped apart by shrapnel, choked by gas, gored by bayonets without losing the will to fight.

German youth was long ago convinced that Nazi destiny is more important than death; French and British youth have found their cause in Hitler's aggressions. But last week as 1,250,000 U.S. students of military age assembled peaceably on the grounds of 1,500 colleges and universities (see *p. 46*), they were still quite sure they had nothing to fight for, and some of them doubted whether any cause was worth the unpleasantness of dying.

Like their elders, whose passions and opinions they reflected, the young men of the U.S. were bewildered by war, undecided how they should react to it. In their campus newspapers they brooded on such problems as encirclement and invasion, debated how the U.S. might be kept neutral. One thing only they agreed on unanimously: they did not want to take up arms in Europe.

Most emphatic undergraduate journal in the East was *The Dartmouth*, only daily newspaper in the town of Hanover, N. H., and a member of the Associated Press. Wrote Editor Thomas Wardell Braden Jr.: "In the last great war men of our age died:

1) for democracy, 2) to crush German Imperialism. These words don't always mean what they say. We need to remember that there are

ideals of truth and realism stronger than the fake ideals which are battering at us from Europe."

"We hardly feel justified," said Editor Braden, "in terming Mr. Roosevelt's party a peace bloc."

Other student papers were more restrained, contented themselves with warnings and prayers. Said the Yale *Daily News:* "Secure from a military and economic standpoint, America will only become involved in the present war if she again heeds propagandist pleas to preserve democracy and stamp out Hitlerism. Let us be on guard against being persuaded to fight for the economic interests of England and France."

The Harvard *Crimson,* under Blair Clark's supervision took its stand with one leg solidly behind the Allies: "The best chance of our remaining neutral is the success of Allied arms." But in the next breath the *Crimson* added: "Americans wishing to remain neutral must make a new resolve to stay out of this war at any price—Allies win or lose."

Ralph Hinchman Cutler Jr., returning as a senior to Harvard after a summer abroad, wrote in the Crimson: "In the present European war there is only one thing at stake: the supremacy and preponderance of the British Empire. The war appears to be merely a clash of rival imperialisms."

The Daily Princetonian had nothing to say editorially about war. But Editor Robert P. Hazlehurst admitted: "There's not much doubt as to how Princeton men feel about the war: we are naturally biased in favor of the Allies." Meanwhile at Vassar College, in the *Miscellany,* Editor Nancy McInerney of South Bend, Ind., spoke for young womanhood: "We don't want our husbands shot. We favor the cashand-carry act because it is more neutral."

Southern colleges were almost unanimous in their sympathy for the Allied cause, but they too leaned toward isolation. Said the University of North Carolina's *Daily Tar Heel:* "We Americans still are for the most part unaware of the ever-engulfing holocaust of screaming bombs, earthy blastings, and flying shrapnel. We are out of range." In the Rollins College *Sandspur* Editor John Henry Buckwalter III exclaimed: "President Roosevelt is absolutely right...." The *Daily Reveille* of Louisiana State University thanked God for 3,000 miles of ocean.

One of a handful of college publications with the bulk and attitude of a metropolitan newspaper is *The Daily Texan,* published by students of the University of Texas at Austin. War news (by International News Service) was played down in the *Texan,* Editor Max B. Skelton noted an increase in student registration, added as an afterthought: "The people of the U.S. should be thankful that the youth of college age are beginning to worry about mathematics and physics instead of watching for the approach of enemy bombers."

At Northwestern University in Chicago, brilliant, political-minded Stanley Frankel, editor of the *Daily Northwestern*, founded a College Front for Peace with the platform: "We will not fight in Europe." He sent letters to editors of 250 other college papers, inviting them to join; at week's end some 50 had accepted.

In the *Northwestern* Stanley Frankel wrote: "The countries of Europe have found an interesting pastime for their youths. They give them guns and airplanes and cannons and bayonets and nice uniforms! We youth in America don't want to play soldier. We want the U.S. to keep out of war."

Like the *Texan*, *The Daily Illini* at the University of Illinois in Urbana is weighty and professional; it leases an AP wire, carries the syndicated Washington *Merry-Go-Round*. Unlike the *Texan*, last week it played the war to the hilt. Editorial Writer Wallace Dooley was judicial, weighed the factors favoring neutrality.

Midwestern students on the whole were wary of U.S. intervention in Europe's war, for any cause.

Stanley A. Frankel

Appendix

Handpicked Articles

As the author has discussed before, he wrote many articles about fighting in the S. Pacific, some of which have been reproduced in the body of this book and also in the appendix. However, he also wrote many magazine articles, mainly non-fiction, unrelated to the war, and several of his favorites have first been published in national magazines and newspapers and then often republished in literary anthologies. Following, in no particular order of event, chronology, or favoritism are those handpicked pieces marked for, the author hopes, some form of posterity.

Stanley A. Frankel

Appendix

Adlai Remembered

by Stanley A. Frankel

Adlai E. Stevenson of Illinois died 25 years ago. This year he would have been 90. We had been friends ever since he ran for Governor in 1948 and I wrote...or, better...tried to write...speeches for him. I can't think of him as a 90 year old; to me, he's forever young...my friend, my boss, and in some activities, my collaborator. I think of that sparkle in his eyes, his warm and gregarious personality, and above all, his finely honed sense of humor...never unkind or cruel...but sharp and always "on the button."

Many incidents come to mind, but the one involving Westchester might be the most relevant. An associate and I called on Governor Stevenson in 1960 shortly after he was named U.S. Ambassador to U. N. We had come up with an idea to put him on TV regularly, a series which did go on thereafter for two years and won a Peabody Award. This was our initial meeting on the subject, an effort to sell the idea to him.

As I was in the middle of my presentation, Miss Roxanne Eberlein, the "Guv's" secretary, rushed in, whispered a few words into his ear, and left. Stevenson apologetically said:

"Stan, the Secretary of State is outside. Would you mind if he interrupted our talk?" We said, quickly, "Of course not ..." and got up to leave the room. With that, Dean Rusk bounced in.

Now, I had known Dean Rusk for many years. He was a Scarsdale neighbor, and we often walked to the railroad station together and participated in many of the same activities. As he walked in and saw me, he was taken aback, for we had not known each other in this political context.

As we shook hands, there flashed through my mind the magnificence of the opportunity: Stevenson and Rusk as a captive audience, at least for a moment, an opportunity to toss the two most powerful statesmen in the U.S. my ideas on Cuba, Viet Nam, the Berlin Wall. But often our tongue does not keep up with our mind; I must confess this is the dialogue: "Dean, remember two years ago when you were program chairman of the Scarsdale Junior High PTA?" "Yes, Stan." "Well, you know the year after, they made me chairman." "Oh yes,… that's right." "Then, last year, when you coached the sixth grade year, I'm the coach." "Oh…I had heard that…nice going…how's the team?"

"Fine…Dean, but this will really throw you. You've been chairman of the Senior High School program committee until you were called away to Washington. Guess whom they selected to fill in for you?" "You mean…you were selected…that is a coincident!" Meanwhile, Stevenson, who had one ear cocked while listening to this scintillating conversation, shook his head and commented sorrowfully: "That's just like Scarsdale

…always jumping from the sublime to the ridiculous!"

Properly, and yet gently, rebuffed

…I prepared to retire inconspicuously from the room…backing away from my seat, toward the door. Unfortunately, Stevenson had a large plastic contour map of the western world on the floor, next to the door, and as I backed away, I inadvertently stepped onto the map. It gave way with a crunch.

Both Stevenson and Rusk quickly walked over, and noticed the broken parts. Stevenson looked up with a humorous gleam in his eye, "Great work, Stan, you've accomplished what we've been trying to do for years

…you've crushed Cuba."

As I mentioned before, we did persuade Stevenson to do the TV series, and it went on, every Sunday, over ABC-TV for two fulfilling years, until he died. Many of the programs were memorable, the ones with Nehru, and Humphrey, and Rusk, all as Stevenson's guests. One I recall, most, the

program aired the Sunday before Christmas, 1962. Stevenson commented on the world situation and his eloquent remarks remain fixed in my mind. The tape of this particular program has been lost, and there was no official transcript. Therefore, I may have missed the exact quotation, but I think I have salvaged the meaning and most of the style.

"It might be a good idea during this Christmas season to remind ourselves about the nature of men…and Man. Men are sometime cruel, but Man is kind. Men are sometime greedy, but Man is generous. Men are mortal, but Man is immortal, and I believe along with Faulkner that Man will do more than survive. He will prevail."

As a grandfather I must believe that the eloquence of my friend, Adlai Stevenson, will not only survive in history…but that his lofty ideals for this nation and the world will eventually prevail.

EDITOR'S NOTE: *The writer, Stanley A. Frankel, has served on Presidential Commissions for the Peace Corps and Youth Opportunity; is President of Phi Beta Kappa Associates; is a senior officer at Ogden Corporation; was a speechwriter for a number of Presidential candidates including Adlai Stevenson, Hubert Humphrey, George McGovern and Robert Kennedy; won a Peabody Award for the Adlai Stevenson TV series; is a distinguished Phi Beta Kappa lecturer, and lives in Scarsdale, N. Y. He's also a much decorated combat infantry man from WWII battles in the South Pacific. And he has served on the N. Y. State Governor's Task Force on High Education, the Chancellor's N. Y. State Panel on Long Range Planning and Remediation; and the Baruch College Ph. D. Board of Visitors.*

Stanley A. Frankel

Appendix

A Not-So-Grand-Fathers' Day*

by Stanley A. Frankel

My Dad died when he was 34. I was 8. His death resulted from a kidney ailment called Bright's Disease. A decade later, Sir Alexander Fleming discovered penicillin; and shortly afterwards, it was learned that one shot of penicillin knocks out Bright's disease in 24 hours.

Etched in my memory is my last visit with him. Dying in the Miami Valley Hospital, Dayton, Ohio, he asked to see me...his only son. My mother and her family debated for some hours the advisability of my going to his oppressive sick room. But, his quiet wish, spoken with slurred speech between bouts of unconsciousness, won the day, and I was brought up to his hospital ward. I walked over to him, noticed how puffy and pale his face was; his eyes were closed and he was still. I gingerly took his hand, and he opened his eyes and tried hard to smile. Not much of one. Then, haltingly, indistinctly, he whispered: "Son, kiss me goodbye."

I knew exactly what that meant, and as I leaned over to give him a peck on the cheek, I broke down, sobbing, buried my head in his shoulder and lay half on top of him as I had done so often in years past. He gently patted my cheeks and ran his fingers through my hair. I heard him say, now clearer than before: "Stanley, son, don't cry for me...be a strong boy

...don't cry, ever."

My mother came to the bed, led me away as he closed his eyes and put his arms alongside his wasted body. I didn't look back as I went out the door. I fought back my tears though I felt a wrenching pain in my stomach.

---

* Reprinted from *N. Y. Daily News*

Two days later he died, and the friends and relatives at the funeral must have been startled by the sight of a stoic little boy, sitting next to his father's casket, looking straight ahead and not expressing one whit of emotion.

Since then, like everyone else, I've had my share of grief and pain, but I have never cried, not even in those moments of exaltation when tears of joy would have been appropriate.

And life has treated me well, including the blessing of an eight year old grandson whose interests and priorities practically coincide with my own. Our first love is baseball, and we play pitch and catch by the hour; we throw a tennis ball against the side of the house for the other to reach; and we bat against the pitching machine at Sportsland. Since he lives fairly close by, we spend lots of time together on our other interests, viewing Phantom of the Opera, going to horror movies, and haunting Yankee Stadium. We chase frogs in Bermuda after dark, wearing the T-shirts his parents gave us: "Frog Hunting Champions of the World;" and when he stays overnight with us, we lie in bed together, watching the goriest TV we can locate, covering up each other's eyes with our hands as the hatchet descends on the unsuspecting victim.

He loves my WWII stories, especially the one about the Japanese soldier in the Solomon Islands who stumbled into my foxhole, fell on top of me, flush against the bayonet I had pointed in the direction of the falling body. At Grandparents Day at his Trinity school, he proudly introduced me to his teacher as "Pa, the solider" pulled a medal I had given him out of his pocket and told her I had won that for the rescue of some American prisoners at Bilibid Prison, Manila. She had me recite the full story to his class while he sat next to me in obvious pride, dangling the medal in front of the class.

We both loved my 89-year-old Uncle Arthur...his great-great Uncle ...and when we visited Uncle Arthur in his Dayton, Ohio nursing home, Adam was responsible for getting Uncle Art out of bed and walking, hand in hand with his great-great nephew around the Home. When we planed back to NY, I asked Adam if he'd like to come back next year to celebrate Uncle Arthur's 90th birthday. Adam replied matter-of-factly "Sure, Pa...but you know that Uncle Arthur isn't going to make it."

Uncle Arthur didn't...and my wife and I flew to Dayton for the funeral. When we returned, Adam was overwhelmed with questions since he had buried his pet hamster only a week before. "Did they put Uncle Arthur in a box? What was he wearing? Did they put the box in a hole? Did they cover the box with dirt? Was everyone sad?"

Adam, I reported to him, this death was not sad because Uncle Arthur was old and very sick. To which Adam responded: "Pa, are you ever going to die?" I explained to him as discreetly as I could that as time goes on, everyone passes on, probably to a better life.

To which Adam began tearing, grabbed me around the neck, and blurted out: "Pa, when you die I want to die, too." And then he sobbed hysterically, burying his little head in my shoulders…as I had done with my Dad sixty years ago.

And…. yes….I cried, too.

… Dad…. please forgive me.

Stanley A. Frankel

Appendix

Wordsmith for Presidential Hopefuls[*]

Stanley Frankel has earned many honors in his life, but he is particularly fond of one that will never hang on any wall—that he made President Nixon's "enemies" list in the early 1970s. For an unreconstructed liberal like Frankel, honorable mention on the Nixon list was gratifying recognition that his political activities had not gone unrecognized. Frankel's speechwriting for such standard-bearers of the liberal faith as Adlai Stevenson, Hubert Humphrey, and Robert Kennedy earned his place on this Dean's List (it was Nixon aide John Dean who revealed the existence of the enemies list at the Watergate hearings). At the same time that he was freelancing as a highly valued speechwriter, Frankel pursued a successful business career and has been teaching at Baruch since the late 1960s.

Now semi-retired at 70, Frankel is hardly slacking off. Although he hasn't done much speechwriting lately, Frankel is an active freelance magazine writer, contributing to such publications as *Good Housekeeping* (he had an article in the December issue) and such local papers as *Town &*

Village (see accompanying article). And he teaches a course on Management in Society two days a week as an adjunct professor at Baruch. "I go for what pleases me," he says simply about his wide range of interests, "but whatever I do, I try to do it with dedication."

What has pleased him about working as a speechwriter is that it put him "in the privileged position" of working with men like Humphrey, Stevenson, and Bobby Kennedy: politicians he revered and who were political kindred spirits. "If you are on the same wave-length as the person you are writing for, it is even possible to get some of your own ideas into a speech," he declares. He has never accepted fees as a speechwriter, believing that this allows him more freedom as a writer.

---

[*] Reprinted from *Baruch Alumni Quarterly*

Starting with Stevenson

In the late 1940s, Frankel worked as a junior editor at *Esquire* in Chicago and first became involved in speechwriting. Adlai Stevenson was running for governor and looking for speechwriters. Frankel was recommended to him as a talented writer with sympatico political views. Frankel would have a long and close association with Stevenson, writing for the Illinois Democrat throughout his political life, while Stevenson, in turn, would further Frankel's speechwriting career by introducing him to Humphrey and others. Frankel later repaid Stevenson this professional debt by becoming executive producer of "Stevenson Reports," a public affairs television program in the 1960s that provided the former UN ambassador with a platform for his ideas during the waning years of his career. The show earned Frankel a prestigious Peabody Award, the industry's recognition for outstanding programming.

Reflecting on his long association with Stevenson, Frankel is most pleased to have been one of those who urged Stevenson to bring the issue of a test ban on atmospheric testing into the public view during Stevenson's unsuccessful bid for president in 1956. At that time, the dangers of fallout and nuclear radiation were not as universally understood as they are now, Frankel explains. During the closing stretch of the campaign, famed nuclear physicist I. Rabi, one of the inventors of the atomic bomb, approached Stevenson, suggesting that he use his influence to push for a ban on atmospheric testing. Stevenson, however, was reluctant at first to embrace the unpopular issue. "I was one of those who pushed Stevenson into really going public with the issue, since I believed that it was a vital one." Roughly two weeks before the election, Stevenson reversed his position and took a strong stand against testing in the atmosphere during a speech crafted by Frankel. Not long afterwards an atmospheric test ban was passed by Congress. "I don't claim any great influence on history but I do feel that I did play a part in persuading Stevenson to start the ball rolling for such a treaty," Frankel says.

On other occasions, Frankel's advice was not followed. In 1968, Frankel tried to persuade presidential contender Hubert Humphrey to take a strong stand against the Vietnam War. With former FCC chairman Newton Minow (Frankel's brother-in-law), Frankel drafted Humphrey's acceptance speech at the notorious Democratic convention in Chicago—including a call for an end to American involvement in Vietnam. "Humphrey liked the speech but was afraid of taking an anti-war position believing that it would anger President Lyndon Johnson," Frankel says. But 10 days before the election Humphrey revised his stand and spoke against the administration position. Frankel wonders "What would have happened during the convention if Humphrey had gone ahead and taken a stand against the war? Perhaps it might have defused some of the violence and atmosphere of confrontation at the convention and changed Humphrey's public image."

One of the Least Rewarding Literary Forms

While political speechwriting has given him a glamorous ringside seat in the political arena, Frankel is less than enthusiastic about speechwriting as a craft. "It has to be one of the least rewarding forms of writing," he declares, "because as an anonymous craftsman you have virtually no ego reward." On the other hand, a good speechwriter is always in demand, he says. "You have to be something of a chameleon, able to take on the voice of many different people." The key, Frankel explains, is to know your subject thoroughly. "You have to know the words with which they are comfortable, their sense of humor, and their speech patterns. For example, you should know whether they favor long or short sentences. You have to really spend time and become totally familiar with their ideas and speech idiom."

Frankel came to speechwriting as an offshoot of his work in journalism and advertising. Graduating from Northwestern in 1940 with a BA, Frankel worked briefly as a police reporter in Chicago and as a public relations writer at the Chicago CBS radio affiliate. Drafted at the outbreak of World War II, a year stint in the army turned into a five-and-one-halfyear [sic] tour of duty in some of the bloodiest action in the South Pacific. Frankel entered the service as a private and came out commanding a battalion, receiving six decorations for bravery—including two Presidential Unit Citations. "I kept getting promoted when all the other officers around me kept getting killed," Frankel says matter-of-factly.

Returning to Chicago after the war, Frankel resumed his career in writing. He was a freelance writer and then junior editor for *Coronet* and *Esquire*. When *Esquire* moved to New York, Frankel also relocated, now as head of the magazine's advertising and promotion division. In 1954, he left the magazine business to become vice president for communications at Ogden, a billion-dollar conglomerate then involved in food and manufacturing products. His over thirty-five years with Ogden have been happy ones, says Frankel, noting that he has been particularly grateful that the company has allowed him leaves of absence during the periods when he worked as a campaign speechwriter. Retired from full-time work with the company, he still spends one day a week as a consultant with Ogden and gives another day as a consultant to the public relations firm of Manning, Selvage & Lee.

A Teacher Who Gives His All

Education has been, along with politics, his other major extracurricular concern. A strong advocate of educational opportunities for the disadvantaged, especially black teenagers, Frankel helped pioneer the YMCA project for youth in Bedford-Stuyvesant. In this program, more than 500 unemployed area youths were given an intensive nine-month vocational training, and the majority of them went on to find employment in their chosen trades. Frankel has also been a member of Governor Carey's Higher Education Task Force and served as assistant chairman of the State University's Long-Range Planning Committee under Governor Rockefeller.

Frankel began teaching at Baruch by filling in for a professor. More than twenty years later, he is still happily at the front of the classroom. He currently teaches a course on Management in Society that examines the place of business in terms of ethics, corporate responsibility, philanthropy, and other issues. "I love teaching at Baruch. But I am most interested in students who don't do well in the beginning. I try extra hard to get them involved with the work so that they eventually do well." His concern for his students is reciprocated in the strong rapport that he has with them: he is always one of the highest-ranked members of his department in student evaluations. He is also generous with his time outside the classroom, meeting with students after class, advising them on their classwork, and helping them with larger questions relating to school and careers.

It is indicative of his interest in these students that has even got his family involved. He notes with some pride that both his sons, who are bank presidents, and his daughter—manager of InFoQuest, the AT& T science and technology museum—have hired his former students. "The rewards of teaching far outweigh the financial return," he says.

Appendix

Teaching at Baruch
*Students' Success Still the Best Test**

by Stanley A. Frankel

Open admissions has been unjustly found guilty of killing the quality of undergraduate learning at, among others, Baruch College, part of the City University of New York (CUNY). My 20 years experience as adjunct professor at the college, located near Gramercy Park at 23rd Street and Lexington Avenue, has proved otherwise to me. A large number of my night and noon-time students have been beneficiaries of New York's open admissions program, whereby any student with a high school diploma is guaranteed a place in some higher education facility in the city.

Now, I can't speak for all of the 16,000 undergraduates studying at Baruch, but I can tell you about that microcosm of 70 who take my course in Business and Society each semester. They range in age from 20 to 60; half of them male; half female, including a heavy proportion of blacks, hispanics and orientals and a small scattering of whites. Most of them work full-time in banks, corporations, accounting firms, taxicabs, and fast food restaurants; plus there's a sprinkling of policemen, firemen, and United Nations employees from various member nations. They go to college from three to 12 hours a week in search of their undergraduate degree in business, which most of them will eventually achieve, not in four years but in five, six or seven. Some of them have difficulty with written and spoken English, and I suspect that most of their high school

---

* Reprinted from *Town & Village*.

205

educations were inferior. But they made up for communications handicaps with motivation; and for inadequate high school preparation with gutsy determination. They substitute street wisdom for book knowledge, and they apply what they learn earnestly and doggedly.

An Example from El Al

An example: ten years ago, a former Israeli pilot, then working on the ground for El Al security at JFK Airport, detained a ready-to-depart El Al flight because the pilot had inadvertently slipped by the sign-in register. The plane was delayed while the furious pilot had to identify himself face-to-face, before the man who was my student. The pilot subsequently filed charges. My student defended his actions thus: "I had the responsibility for making sure that the pilot was the pilot. I felt it was my obligation to postpone the take-off until I was sure. My upside risk was a one-hour departure delay; my downside risk was a possible hijacking." In the hearing, that student was not only exonerated but praised, and one year later was promoted to security chief. He had thoroughly learned what our textbook taught—that authority and responsibility in business management are inseparable. In fact, he brought his textbook and class notes into his hearing, liberally quoting from both.

Great Strides

Most of these students have made great strides, some from a very low base. My own definition of higher education is simple, maybe even simplistic: "Higher education means higher when you finish the course than when you began."

I recall the Japanese youngster who labored over the text and the lectures. I provided him with my lecture notes because he couldn't keep up with my classroom verbalizing; and I advised him to carry a pocket radio with him at all times—tuned into news programs. He earned a B in the course, with an 80 on the final, after a 40 on the first midterm. The last third of the semester he was taking his own lecture notes, walking into and out of class with the radio pressed against his ear.

Quotes Dr. Salk

My first lecture always begins with a line from Dr. Jonas Salk: "Failure is not, not succeeding; failure is not trying," and I manage to mention that line at least once a month each session. I persist in reminding them that "Nothing in the world can take the place of persistence. Talent will not: nothing is more common than unsuccessful men with talent; genius will not; unrewarded genius is almost a proverb; education alone, will not; the world is full of educated derelicts. Persistence alone is omnipotent."

Coming out of often horrendous high school backgrounds and deprived home life, they arrive with motivation, and I keep fueling that fire by quoting Dr. Benjamin Mays: "The tragedy of life does not lie in not reaching your goals. The tragedy lies in having no goals to reach. It isn't a calamity to die with dreams unfulfilled; but it is a calamity not to dream. It is not a disgrace to fall short of getting to those stars; but it is a disgrace not to have stars to grasp for. Not failure, but low aim is a sin."

In my first-day course expectation handout I explain that if they attend all the classes, are punctual and alert, do their homework assignment, read their textbooks and listen to the lecture, there is no way they won't do well in my course. If they do all of these things and fail one of my exams, it is not student failure but teacher failure, for I haven't been able to reach them. I assure them that I don't intend to goof and they don't either. And they don't!

Motivation Is Key

Some of them are quicker than others; many of the foreign-born students have great difficulty with our language, and I've even had a few who came into my class reading at a sixth grade level. But I have found that the brighter, better prepared students aren't slowed down by those at the other end of the intellectual scale, for many of those who got into Baruch through open admissions possess superior drive, desire, and motivation. No one forced them to go to college but themselves. These qualities—persistence, determination, a moving kind of reverence for the professor, the textbook, the college—enrich my class far beyond the dimensions of high IQ and inherited cultural backgrounds.

Since I, too, am part-time (a corporate officer eight hours a day and an adjunct professor noons, evenings, and weekends), I derive a rich reward of tremendous psychic income from my city university gang, while I earn a monetary return from teaching, on an hourly basis (if all classroom, transportation, grading, homework reading, and lecture preparation time are factored in), very close to the legal minimum wage.

But where else would we adjuncts receive such a spiritual lift than from these eager young-and-often-older learners, who are there every day or night to absorb as much as they can in spite of missed meals, four-hours-anight [sic] sleep, crowded subway trains and the unkindest cut of all—the charge that they are responsible for a deteriorating quality of education.

I love those hundreds upon hundred who have come to me but have not really gone; who write me thank-you notes which make my wife cry; who invite us to their weddings; who have moved up in their business careers and call on me to recommend my current students so they can hire them.

And if I had a modicum of religion, I would be thanking God every night for my golden opportunity—and the open admissions plan which made these golden hours and years possible—for all of us.

EDITOR'S NOTE: *The term "open admissions" a used in Professor Frankel's essay refers to a policy of the City University that provides a place in a CUNY institution to graduates of New York high schools. Students with a grade point average (GPA) below 80 in academic subjects are admitted to a two-year college in CUNY. From there, they can transfer to a bachelor degree-granting senior college such as Baruch if their associate degree performance meets transfer standards. In 1990, admission to Baruch as a freshman requires a GPA of 83 or combined SAT score of 990. In an average year, only 50 percent of the freshmen and 75 percent of the transfer applicants are granted admission.*

Appendix

If I Were 21[*]

by Adlai E. Stevenson with Stanley A. Frankel

I have often wondered what magic lies in the age of 21. The day before our 21st birthday, we are considered immature, uninformed and not responsible. Then suddenly, a strange alchemy remakes our legal and moral selves: overnight we become independent, self-sustaining and competent citizens of the Republic.

One day, we are, for all practical and lawful purposes, children. The next, we select presidents, send men to jail, and sometimes inherit the right to squander money which, until now, has been prudently denied us.

Whatever it is—the 21st birthday is about as decisive and pivotal a 24-hour spate of time as any man is apt to have in his life.

Actually, we all know that 21 is no more than an arbitrary, imaginary equator marking off youth from manhood and womanhood. Society said long ago: there has to be *some* point at which to refurbish voting lists and cut umbilical cords—and 21 seemed to be a happy figure. And I suspect that it was selected by solemn, elderly gentlemen profoundly mistrustful of radical, impetuous youth, to whom anything younger than 21 would be risky indeed.

In my case, however, I cannot recall that I was impressed by the significance of this magical age. To be sure there was hilarity and the 21-candle cake. There was my diploma—in sight at last. And there was the privilege of voting. There was also the sudden opening of a Pandora's Box of decisions: would I teach, be a reporter, a rancher, study law....

---

[*] Reprinted from *Coronet Magazine*, *The N. Y. Post Syndicate*, and a number of anthologies.

And while the prospect of earning a living and supporting a family must have been sobering, I can't remember feeling any acute anxiety about the future or doubting my adequacy to meet whatever challenges the years would bring.

It wasn't long though before I skidded to a tentative stop, chastened by the realization that all of the regalia of maturity I had acquired was largely symbolic. How very unfortunate, I now chide myself, that 21 had to be wasted on me when I was so young.

Yet—what do I know now that I didn't know at 21?

Whatever it is, as I once tried to put it, it is for the most part incommunicable: "The laws, the aphorisms, the generalizations, the universal truths, the parables and the old saws—all the observations about life which can be communicated readily in handy, verbal packages—are as well known to a man at 21 as at 55. He has been told them all, he has read them all, and he has probably repeated them all—but he has not *lived* them all.

"What he knows at middle age that he did not know when he came of age boils down to something like this:

"The knowledge he has acquired with age is not a knowledge of formulas, or forms of words, but of people, places, actions—a knowledge not gained by words, but by touch, sight, sound, victories, failure, sleeplessness, devotion, love—the human experiences and emotions of this earth and of one's self and other men. Perhaps, too, a little faith, a little reverence for things you cannot see."

Yes, there are things I would do differently if I were on that equatorlike [sic] dividing line of 21 again. I think I like to think, that rather than breathing a sigh of relief at blessed release from classrooms, I would begin educating myself, in earnest. I would rediscover the nearest library—and many of the books I had glanced through with one eye on the report card and the other on the next game. I would try learning, for learning's sake—not for my diploma's—or my parents'—or my ego's.

I would look hard for the inner meaning of the great classics instead of playing a guessing game with my examination questions. I would read, read, read. I would soar where curiosity took me, not just where the recommended reading list pointed. I would be guided by a hungry mind, not by the instinct of competition and survival. And I would question—question everything.

Looking back, I feel that, more than by any other single factor, imaginative, healthy youth is characterized by rebelliousness. It's a good thing, and normal to inquiring youth's uncorrupted vision of pure justice and goodness. It is good for man at every age to seek, to question, to rebel—to keep alive and up to date in body, mind and spirit. Change and progress are the fruit of our re-examination of the methods, attitudes and customs we have taken for granted; they are the fruit of rebellion and rejection of the old.

Our century cries out for boldness, imagination, experiment—for people, as I have said before, "who take open eyes and open minds into the society they inherit."

But in the impetuous rebellion of youth against all the evils that the children of God have contrived, I would go slow. Of course our 21-year-old need not, must not, swallow whole all the tribal beliefs, modes, manners which have been poured on us by parents, teachers and friends. But neither would I automatically throw out whatever I had been told to accept on faith—whatever didn't yield a simple satisfactory explanation to superficial study. I would try to keep 21 the age of the suspended final judgment, the re-evaluation of our moral and political environment.

As a matter of fact, I don't think my generation at 21 rebelled against much of anything. We were just emerging from the first world war and we thought we were on the threshold of everlasting peace and prosperity. It was the age of "flappers," cynical materialism—and normalcy!

But if not of rebellion, it was a period, I think, of irreverence: there was too little of God and the eternal verities in the air when I was 21 and too much talk just of getting a job, making money, somehow, anyhow, and having a good time. It was smash and grab, and devil take the hindmost.

Today, at 21, I would try a little harder and a little sooner to understand that it is not public demonstrations of reverence but the content of religious convictions that really matters; that there are absolutes of religion and morality by which we shall be judged; and that we need God all the time, not just when we are in trouble.

There are so many things I would do if I were 21 again, or at least *should* do! I would, for instance, participate actively in the political life of my community, my neighborhood, my block. How easy it is to look down disdainful 21-year-old noses at politics and politicians! But that is to default in the basic, never-ending fight for democracy. Far better to get to work in the political party of our choice—to let rebellion and reform do battle in the arena, not the grandstand.

If we are prepared to fight and to die or our democratic ideals when they are threatened from without, why not fight and live for them when they are threatened from within? They always are. And the basic struggle takes place every day, and in your own town.

That is why, since 21, I have learned never to underestimate the precinct captain. He is more effective, in his field, for good or for ill, than a July 4th political orator who throws back at a noise-deaf crowd the platitudes it wants to hear. There is no more eloquent expression of democracy than a sincere man persuading his next-door neighbor to vote for his alderman.

I now know that the most elemental expression of our belief in democracy is exercising our right to vote. A genuinely free and an honestly informed people will ultimately triumph over intolerance, injustice and evil from without or within. But a lazy people, an apathetic people, an uninformed people, or a people too proud for politics, is not free. And it may quickly be a mob.

While paying deference in this atomic age of infinitely complex problems to the specialist and the technician, I would avoid an easy acceptance of another's thinking.

After all, the great issues of the day are not technical, they are moral. And in a thriving, full-bodied democracy, the moral issues are best decided by a consensus which can only evolve when people—and I mean all the people—reason together, reason reason [sic] their way to clarity of judgment and unity of purpose. How often we have observed the great body of public opinion slowly, clumsily perhaps, arrive at moral decisions which are wiser than those reached swiftly, smoothly by specialists—or computing machines.

And, speaking of specialization, at 21 I would not take any job just because the pay is good or the practical prospects bright. The world's work is vast. Each man who labors at his own job to his best ability, happy in his work, has a dignity that cannot be classified. There is no second-class citizen—or worker—in our great nation. The artisan stands equal to the judge; and the truck driver's contribution to a free, strong nation is as indispensable as the comptroller's.

Einstein once wrote that if he had it to do over again, he would have been a plumber. How much better off many of us—less gifted than Einstein—would really be if we resisted the snobbish temptation to take whitecollar [sic] work and followed instead a natural bent to work with our hands and muscles! There is incomparable satisfaction in building, repairing, conserving, producing with our hands. It brings most of the beauty and utility in the world. And how much happier many people

Appendix

would be to go home at night with dirt under their fingernails instead of inkstains [sic] on their fingers, tired instead of nervous!

No matter what job I took at 21, I would not go into it with the conviction that it would be my last. I would not be afraid to experiment in the search for satisfaction. And while I know how hard it is, I would dare to take on bigger assignments than I was sure I could handle, and I would try to work for bigger, better men than I.

To trade integrity for a quick promotion or to sacrifice self-respect and conviction for the boss' favor is a price I would not pay. Better to be fired for the right cause than to sell your talents for the wrong one. You won't have an opportunity to *try out* your idea and ideals, unless you resist the temptation to *sell them out. Conscience* is a fragile thing. It dies easily but the pain lasts forever. You have to live with yourself, and hypocrisy is an uncomfortable companion.

If I were 21 I would try a variety of things on the side to see where my interests led me. I would always seek an hobby quite different from my work.

For health and well-being I would also take up a sport. Even if our participation is crude, even embarrassing, there is more health and physical satisfaction in playing games badly than in watching professionals play them well. And I say (with selfconscious [sic] concern) that I fear there may be some correlation between the fat that accumulates around our middle and the fat that invades our heads.

So, in my recreation as well as my work, I would start at the beginning of adulthood to develop the whole me, with an aim at perfection but an understanding that the aim, not the achievement, is the important thing.

If I were 21 I would hope for a prompt realization that doing for others is not only a Christian obligation, but also life's greatest satisfaction. A neighborhood boy's club would especially interest me at 21 because too many of us get interested in juveniles, not to prevent delinquency but because of delinquency. Our interest comes too late.

Most young men nowadays find their lives interrupted by several years of military service. It seems to me that a young man who fully understands that each generation must pay a price for the freedom to make its choices would accept this duty with enthusiastic loyalty and eagerness to make the most of it. I would learn the soldiers' or sailors' or airmen's trade, and seize this chance to make new friends among men of widely varying interests and beliefs. I would study with fresh curiosity the new places I saw, nor overlook the opportunities for education and skills which the services offer. I would wear my country's uniform with pride and try to bow gladly to discipline in the knowledge that a team is often more important than an individual player. Our greatest batters have to know how to lay down a sacrifice bunt.

Growing up in this Age of Anxiety, the Age of the Hydrogen Bomb and international hysteria, I would expect of my country's leaders good sense, maturity and consistency in dealings with friends and enemies alike. I would accept in good faith the proposition that while all the ordinary peoples of the world want peace and a better life, the aims and methods of the Western and Communist leaders differ widely. And I would also try to remember that no other people have as much as we do: that misery, ignorance and desire still afflict much of the world, and that we dare not lower our guard while working for the peace and well-being of all mankind, regardless of race, color or geography.

Finally, and most of all, I would try to understand, to know, to feel, the hopes and fears of my contemporaries rich and poor, from town and country, that I might better share and influence my generation—a generation destined to live in an exciting, perilous golden age.

There is nothing so fine as to be 21 and an American. One is for a fleeting instant—and the other is forever. So live—decently, fearlessly, joyously—and don't forget that in the long run it is not the years in your life but the life in your years that counts!

Rudolph that Amazing Reindeer*

by Stanley Frankel

His lovable antics have delighted millions of children; here is the inspiring story of how he was born when a father tried to comfort an unhappy little girl.

On a december [sic] night in Chicago ten years ago, a little girl climbed onto her father's lap and asked a question. It was a simple question, asked in childish curiosity, yet it had a heart-rending effect on Robert May.

"Daddy," four-year-old Barbara May asked, "why isn't my Mommy just like everybody else's mommy?"

Bob May stole a glance across his shabby two-room apartment. On a couch lay his young wife, Evelyn, racked with cancer. For two years she had been bedridden; for two years, all Bob's small income and smaller savings had gone to pay for treatments and medicines.

The terrible ordeal already had shattered two adult lives. Now, May suddenly realized, the happiness of his growing daughter was also in jeopardy. As he ran his fingers through Barbara's hair, he groped for some satisfactory answer to her question.

For Bob May knew only too well what it meant to be "different." As a child he had been weak and delicate. With the innocent cruelty of children, his playmates had continually goaded the stunted, skinny lad to tears. Later at Dartmouth, from which he was graduated in 1926, Bob May was so small that he was always being mistaken for someone's "little brother."

---

* Reprinted from *Good Housekeeping Magazine*.

Nor was his adult life much happier. Unlike many of his classmates who floated from college into plush jobs, Bob became a lowly copy writer for a New York department store. Later, in 1935, he went to work writing copy for Montgomery Ward, the big Chicago mail-order house. Now, at 33, May was deep in debt, depressed and miserable.

Although Bob didn't know it at the time, the answer he gave the touslehaired [sic] child on his lap was to catapult him to fame and fortune. It was also to bring joy to countless thousands of children like his own Barbara. On that December night in the shabby Chicago apartment, May cradled the little girl's head against his shoulder and began to tell a story ...

Once upon a time there was a reindeer named Rudolph—the only reindeer in the whole world that had a big red nose. Naturally, people called him "Rudolph the Red-Nosed Reindeer." As Bob went on to tell about Rudolph, he tried desperately to communicate to Barbara the knowledge that, even though some creatures of God are strange and different, they often enjoy the miraculous power to make others happy.

Rudolph, Bob explained, was terribly embarrassed by his unique nose. Other reindeer laughed at him; his mother and father and sisters and brothers were mortified too. Even Rudolph wallowed in self-pity.

"Why was I born with such a terrible nose?" he cried.

Well, continued Bob, one Christmas Eve, Santa Claus got his team of four husky reindeer—Dasher, Dancer, Prancer and Vixen—ready for their yearly round-the-world trip. The entire reindeer community assembled to cheer these great heroes on their way. But a terrible fog engulfed the earth that evening, and Santa knew that the mist was so thick he wouldn't be able to find any chimneys.

Suddenly Rudolph appeared—his red nose glowing brighter than ever—and Santa sensed at once that here was the answer to his perplexing problem. He led Rudolph to the front of the sleigh, fastened the harness and climbed in. They were off! Rudolph guided Santa safely to every chimney that night. Rain and fog—snow and sleet—nothing bothered Rudolph, for his bright nose penetrated the mist like a beacon.

And so it was that Rudolph became the most famous and beloved of all reindeer. The huge red nose he once hid in shame was now the envy of every buck and doe in the reindeer world. Santa Claus told everyone that Rudolph had saved the day—and from that Christmas Eve onward, Rudolph has been living serenely and happily.

... Little Barbara laughed with glee when her father finished. Every night she begged him to repeat the tale—until finally Bob could rattle it off in his sleep. Then, as Christmas neared, he decided to make the story into a poem like "The Night Before Christmas"—and prepare it in booklet form, illustrated with crude pictures, for Barbara's personal gift.

Night after night, Bob worked on the verses after Barbara had gone to bed, polishing each phrase and sentence. He was determined his daughter should have a worth-while gift, even though he could not afford to buy one.

Then, as May was about to put the finishing touches on "Rudolph," tragedy struck. Evelyn May died. Bob, his hopes crushed, turned to Barbara as his chief comfort. Yet despite his grief, he sat at his desk in the quiet, now-lonely apartment, and worked on "Rudolph" with tears in his eyes.

Shortly after Barbara had cried with joy over his handmade gift on Christmas morning, Bob was asked to an employees' holiday party at Montgomery Ward's. He didn't want to go, but his office associates insisted. When Bob finally agreed, he took with him the poem—and read it to the crowd. At first the noisy throng listened in laughing gaiety. Then they became silent—and at the end, broke into spontaneous applause.

Several Ward executives asked Bob for copies. Then someone suggested: why not put the poem into booklet form as a free gift of Ward customers the following Christmas? Next year, 1939—a year in which Bob labored to pay his debts and keep Barbara fed and clothed—2,400,000 copies of the book were printed and given free to youngsters whose parents were customers at the hundreds of Montgomery-Ward stores all over the country.

The story of the reindeer caught on immediately. Psychologists, teachers and parents hailed Rudolph as a perfect gift for children. Newspapers and magazines printed stories about the new hero. Ward's stores and catalogue offices, placing orders for the following Christmas, asked for 3,000,000 copies.

Meanwhile, May won acclaim—but little else. Montgomery Ward owned the copyright. Yet May was happy in the knowledge that his child—and millions of other children—loved his red-nosed reindeer.

Then the war came, and the giveaway project was shelved. Throughout the war years, however, requests poured in for Rudolph books, toys, games, puzzles, records—all nonexistent. And the demand mounted each

Christmas season as parents got out the old Rudolph book and read it to growing families of new Rudolph enthusiasts.

Meanwhile, Rudolph's success did things to Bob May. He forgot his pessimism, began to laugh again and associate with friends. And among those friends was a pretty brunette, a secretary at Montgomery Ward's. In 1941, Bob married Virginia Newton. Together they created three new Rudolph fans—Joanna, Christopher and Ginger.

Finally the war was over—and Ward executives planned a new Rudolph book for Christmas, 1946. More than that, a message came from Sewell Avery, president of Ward's. Touched by the beauty and simplicity of the Rudolph story, he ordered the copyright turned over to Bob—so that May could receive all royalties.

In 1946, 3,600,000 Rudolph booklets had been distributed by Ward's. Promptly a deluge of demands for Rudolph products swamped Ward's and Bob May. Businessmen wired, telephoned and called, seeking permission to manufacture toys, puzzles, slippers, skirts, jewelry and lamps.

A special recording of the poem was made by Victor. Maxton Publishers, Inc., bought the rights to produce a bookstore edition in 1947. Parker Brothers brought out a Rudolph game. Even Ringling Brothers-Barnum and Bailey circus proudly exhibited a pony, equipped with antlers and an electrically lighted red nose, called "Rudolph the Reindeer."

Christmas of 1947 was the brightest ever for Bob May, his family and Rudolph. Some 6,000,000 copies of the booklet had been given away or sold—making Rudolph one of the most widely distributed books in the world. The demand for Rudolph sponsored products increased so much in variety and number that educators and historians predicted Rudolph would come to occupy a permanent niche in the Christmas legend.

Sellouts all over the country inspired merchants to make even more elaborate plans for Christmas, 1948. A special feature is the cartoon in Technicolor directed by Max Fleischer and narrated by Paul Wing which is being run this Christmas season in thousands of film houses. Manufacturers are already blueprinting Rudolph merchandise for 1949-1950-1951—with each item sold returning a royalty to Bob May.

His fortune has now been made, and the years ahead look even brighter. Today, Bob is still a shy, thin, affable man who wants more than anything else to build security for himself and his family. He still works at Ward's—now as retail copy chief—and tackles the job with the same perseverance which has characterized his whole life.

Through his years of unhappiness, the tragedy of his first wife's death and his ultimate success with Rudolph, Bob May has captured a sense of serenity. And as each Christmas rolls around, he recalls with thankfulness the night when his daughter Barbara's question inspired him to write the poem that closes on these lines: But Rudolph was bashful, *despite being* a hero!

And tired! (His *sleep* on the trip totaled zero.)

So that's why his speech was quite short, and not bright—"Merry Christmas to all, and to all a good night!"

Stanley A. Frankel

The Tragic Truth About Our Jury System[*]

by Judge Julius H. Miner with Stanley A. Frankel

Some day you may face a jury, perhaps accused of a crime you did not commit. Or you may be serving as a juror, sworn to pass on the innocence or guilt of one of your fellow citizens.

In the first case, could you expect a fair and impartial trial? In the second case, would you know how to administer that same fair and impartial justice to the accused?

As a judge of several years' experience in the Chicago Circuit Court of Appeals, my reluctant but honest answer to both these questions is "No!"

For instance, it is on record that a citizen of Bangor, Maine, sat on a jury for days before it was discovered that he was stone deaf. And the judge of an Arkansas municipal court reports that one jury foreman in his court read a verdict thusly: "We, the jury, find *our client*, the defendant, not guilty."

In an important Eastern case, the jury voted six for conviction and six for acquittal. Rather than debate the issue for two or three days, the jurors decided to draw a number between one and 100. The juror whose age came closest to that number would make the decision for all.

Recently I presided over a murder trial in which both prosecution and defense refused to call the only eyewitness to the crime. Both lawyers feared that he might unduly influence the jury. In other words, both sides were unwilling to trust the jury with the truth.

---

[*] Reprinted from *Coronet Magazine*.

Daniel Webster said: "Justice is the great interest of man on earth." And in pursuing his great interest, man has established the jury system as a special safeguard against injustice. Yet today, trial by jury is a farce…a mockery of justice…a thing frequently of "sound and fury, signifying nothing."

Why has trial by jury fallen so low? Part of the answer lies in the low caliber of the jurors themselves. The men and women chosen to listen and weigh and consider are generally unqualified for the task. In fact, many prospective jurors would be rejected for ordinary duties by private employers. Why foist them on a court and expect intelligent results? National and state laws specifically exempt many citizens ideally suited for this work. Doctors, government officials, clergymen, lawyers, schoolteachers, newspapermen and many others are exempt from jury duty. Further, our "indispensable" men who don't relish a few dollars a day for jury service usually present perfectly legitimate reasons for being excused.

Getting an eligible man excused on some pretext has become an invaluable asset of small-time politicians. Judge Robert Stewart Sutliffe, famed jurist of New York City, wrote: "Jurors are a lot of men picked from poll lists who have not enough political pull to get off, or who are out of a job and want to pick up a few dollars a day."

If by chance an intelligent group of men and women are "stuck" with jury duty, even then justice and truth are elusive. Is it reasonable to expect 12 untutored, legally naïve jurors, straight from the kitchen, shop or office, to deal wisely with involved principles and technicalities of the law? It would be just as foolish to summon a judge or lawyer to an airplane factory to decide a problem in aerodynamics.

In one case, after five days of trial of a desperate gunman, the jurors passed a note to my bailiff: "What does the Judge mean when he says 'sustained' and 'overruled'?" Unfortunately, this amazing ignorance of the basic vocabulary of a low court is common.

One judge instructing a jury said: "If you find the defendant did, with malice aforethought, project, propel, and/ or otherwise with force or violence, insinuate the aforesaid bullet in, on, against, and within the body of the corpus delicti, then you must bring in a verdict of guilty."

In reply, one of the baffled jurors blurted: "Okay, Judge, but what if we just find that the guy sitting over there shot and killed the other guy who ain't here?"

It is downright scandalous to expose 12 well-meaning but naïve citizens to sharp, high-powered, battlescarred [sic] lawyers who are masters at the art of appealing to human sympathies and prejudices. Under "solemn obligation" to their clients, these shrewd barristers in a criminal case will hammer away at such things as "reasonable doubt," "moral certainty" and "presumption of innocence" until there is unconditional mental surrender by even the hardiest juror.

Not long ago a jury awarded $10,000 to the plaintiff, but the decision was such an obvious miscarriage of justice that the judge set the verdict aside and admonished the jury. When asked how they arrived at such a patently ridiculous decision, one juror replied frankly:

"We couldn't make head or tail of the case, or follow all the messing around the lawyers did. None of us believed the witnesses on either side, anyway, so we just made up our minds to *disregard the evidence on both sides and decide the case on its merits.*"

Court-wise defense lawyers follow a pattern, aimed at confusing the inept jurors by beclouding the real issues. They assail the police as vicious and corrupt; they ridicule the state's attorney as an ambitious tyrant trying to make a record of convictions for political advantage; they imply that if the defendant is found guilty, the judge, an unmitigated sadist, will give the poor, wronged angel the maximum allowable punishment.

Continually I am amazed at how this "underdog" technique works, even if the defendant is guilty beyond all doubt. Finally the attorneys wind up waving the flag, crying that to do otherwise than return this man to his wife and children would be to negate the American guarantee of life, liberty and the pursuit of happiness.

A wife, a child, or an aged parent is always an asset to a man charged with anything from picking pockets to murder. Pretty faces, shapely figures, hospital cots and crutches are flaunted.

One lawyer asked the jury to look at the defendant's white-haired mother, sitting up front in the courtroom. This angelic little lady won the day for her son, for no jury would ever believe that she could possibly have given birth to a murderer. "Mother" was later discovered to be a waitress in an underworld night club.

Defense attorneys prefer their clients to be tried during the Christmas season, when the quality of mercy is particularly unstrained. Jurors have a way of passing out Christmas gifts of freedom or light sentences to criminals who at other times would be sent up for years. And if one of the jurors has a birthday during the trial, his fellow-jurors help him to celebrate by going easy on the defendant.

In one trial, an expectant father succeeded in getting a fast acquittal for an obviously guilty man because the other jurors agreed that "it wasn't fair" to keep the anxious father-to-be waiting around while they tried to reach a verdict.

One of my colleagues tells of a jury composed of many nationalities, a most sentimental jury which in an entire term of listening to cases hadn't returned a single verdict of guilty. Finally an Italian appeared before them for trial, accused of grand larceny, and the evidence against him was so conclusive that the jury voted 11 to 1 to send him up.

The lone holdout was a fellow Italian who spoke up bitterly: "You have been acquitting Irishmen, Germans and Jews right along. Now an Italian comes along and you want to send him away. No!" The 11 abashed jurors hastily rectified their "injustice."

The law that all 12 jurors must reach unanimous decision has produced a bookful [sic] of "unusual" pronouncements from all parts of the U.S. In Easton, Pennsylvania, an impatient judge who had been waiting for a verdict warned the jury that if they took much longer he would lock them up overnight. The jurors quickly reached the verdict, but it developed later that two of them, unable to make up their minds, had decided the issue by tossing a coin.

Some lawyers and even judges contend that the answer to the abuses of jury trial is to eliminate juries altogether. That is out of the question. The system is too deep-rooted in our American philosophy of life and law. It is the greatest guarantee for a fair trial yet conceived, and no one who truly believes in a free society should advocate its abolition or the diminution of its power.

Yet there are certain commonsense changes in the system that would alleviate most of the abuses I have pointed out. Here is a list of minor reforms that I believe would reimplement the jury as an instrument of justice:

1. Abandon the requirement for unanimous agreement of 12 jurors to reach a verdict. A majority of one in our electoral college elects a President: the majority principle determines the outcome of all elections. And yet it requires a unanimous verdict of 12 to convict or acquit a moron or vicious gunman. This rule of unanimity is nothing less than legalized coercion. It condemns an honest difference of opinion; it makes jury service disagreeable; it is an incentive to corrupting jurors.

2. Judges should be empowered to strip courtroom procedure of its confusing side issues, keep the language understandable and restrict lawyers in their befuddling techniques.

3. Judges should further be empowered to assist in and accelerate the selection of jurors, while the practice of lawyers to resolve on the least intelligent, instead of the most intelligent, jurors should be curtailed.

4. Jury duty should be broadened to include many classes of citizens now excused from serving. Furthermore, the ease with which men escape jury duty should be stopped. Jury service is a sober civic responsibility, and every American citizen should be required to take his turn.

5. Lastly, and most important, all jurors should be required to undergo a short, intensive course in legal terminology, procedure, obligations and duties. This school should be administered by neutral representatives of the bar association, law schools or the courts, and should cover points which have relevancy to a trial. In fact, a brief course in jury service could easily be conducted in our public schools. After all, we train dispensers of liquor behind the bar. Why shouldn't we also train dispensers of justice before the bar?

Chief Justice John Marshall once wrote: "The judicial department comes home in its effects to every man's fireside; it passes on his property, his reputation, his life, his all." The jury is an indispensable part of that judicial department. But unless it functions efficiently, our liberty is insecure and the administration of justice a mockery.

If justice is not to be found in the courts, then the American way of life is in jeopardy. A few minor reforms in our jury system will speedily remove much of the danger, and perform for our citizenry a good that is far in excess of the slight energy required to effect the changes that experience and common sense dictate.

Stanley A. Frankel

Appendix

A Baseball Memoir*

by Stanley A. Frankel

Behind my desk is a fading picture of the Northwestern University freshman baseball team of 1937. Though I'm one of the smaller members, the photographer placed me in the back row, so the bottom of my uniform didn't show. I was issued a team shirt but they ran out of baseball pants my size so I was wearing sweat pants, which, in the interest of harmony, were shielded from the camera.

I worshiped baseball, loved pitching, and was perfectly miscast for an activist role. My build was slim and my hands were grotesquely small. But I had a passionate desire and plenty of practice time. I devoted my high school summers to pitching American Legion baseball in Peoria, Ill., where I visited a rich and lonely cousin. Without any muscle power, I developed an assortment of junk pitches, curves, and knuckleballs and those, plus exquisite control, got me by.

When I tried out for the freshman team at Northwestern, the coach, Maury Kent, an old Brooklyn Dodger utility infielder, sized me up quickly; he watched me throw a few not-so-fast balls and decided against spending the five dollars for the lower half of my uniform.

For two weeks, I came to practice, shagged flies, and never pitched, not even in batting practice. I became desperate for a chance to exhibit my small and slow talents.

I found the angle. As a *Daily Northwestern* reporter, I befriended the sports editor who asked me, in response to some strong hints, to write a piece about NU's freshman baseball prospects, *sans* byline. So...I drafted a lengthy analysis, half the verbiage on the pitchers. Ignoring several

---

* Reprinted from *Northwestern Perspective*, September 1989.

giants with impeccable credentials, I laid it on heavy when it came to "the leading prospect, Stanley Frankel, who curveballed his way to an American Legion championship last year." Hedging my flowers a bit, I concluded, "If the sore arm that Frankel picked up last summer is cured, he should be one of the great Northwestern twirlers of the decade."

The story appeared Tuesday morning, and I went out for practice that afternoon. The coach called roll, came to my name, looked up, smiled broadly, and inquired why I wasn't in full uniform. I explained they had run out of trousers, so he yelled at the manager to outfit me then and there because "Frankel is starting against the varsity today."

The pants were too long and I was nervous, but I did get the call and threw my weird assortment of slow junk against a varsity team that had been fed straight, wild, and fast balls for two weeks. In two glorious innings, no one got a clean hit although a couple of squibblers were beaten out because the freshmen had a leadfooted [sic] third baseman. I was happy and proud, and when Kent came to the mound in the third inning, I figured I was in for some roses. Instead, looking sad, he put his arm around me and pronounced the verdict: "Frankel, I can tell that sore arm is killing you and it takes tremendous pain and effort for you to get the ball to the plate. Better take it easy for the rest of the season and save that arm for next year. Or better yet, take up golf."

He broke my heart and, of course, I quit the team. The anti-climax was even more humiliating. Two years later, a close friend of mine, Stan Klores, was named freshman coach. Stan attended college one semester a year, and the other semester he was on a Chicago Cubs minor league team and labeled one of the better outfield prospects in the Cubs' organization. He knew my tragic tale, and over a chocolate soda at the Goodrich fountain, he wondered aloud if I'd like to come out again, my senior year. I was thrilled, but then remembered he was a freshman coach. What would I—an aging senior—be doing with the freshman team? "Well," Stan said, halfwinking [sic], "I'm a great believer in psychology. If you pitch batting practice to my sluggers, they'll hit you so hard that it will give them confidence."

Some joke.

Appendix

Thoughts on the Fourth, and the Fourteenth, and...*

by Stanley A. Frankel

I was recently invited by Baruch College Center for Management to address a group of visiting Chinese graduate students sent by their government to study American methods of management. The center had an intensive two-week schedule for working with these young people, and my assignment was to talk about corporate communications, a vineyard in which I've labored for almost a half-century.

But events of the past several weeks in China and elsewhere in the world have reduced my assignment to trivia and irrelevance. Instead, I will lecture to them...and listen to them

...on the meaning of the present Chinese Revolution.

In shorthand, the meaning is obviously freedom, and I must discuss with them the freedom that I know

...the freedom I've studied and written about...and the freedom I fought for in World War II as a combat infantry officer in the South Pacific.

I feel I must remind them that, for all good intentions, freedom is not an end, but a means, and historically, some revolutions, such as the one the

French celebrate on July 14, begin in idealism and glory and end up in abuse and betrayal...and a net loss in the freedom for which so much blood was shed and treasure spent.

The concept of freedom itself is as old as man. Theologians tell us that God, in creating man, endowed him with freedom which, in its essence, was the priceless gift of the ability to choose between good and

---

* Reprinted from *Westchester Spotlight*, July 1989.

evil. Freedom is a beautiful word, which has at times been put to ugly use; it is an adult thought which has often been adulterated; it has a ringing sound, but it is occasionally wrung dry by some who act as if they had discovered it.

Our own founding fathers proclaimed "Let freedom ring," but they did not contend that they had invented the idea. They *perfected* it. The great genius of American spirit, behind our Revolution, was the recognition that the cause of freedom is the cause of the individual. This individuality, this God-given right to choose, is the basis of practically every value in human life—spiritual, moral, intellectual and creative. Our freedom means the right to one's dignity, and in a free society, no individual or group of individuals being entitled to diminish the dignity of another. Indeed, our Revolution was unfinished until the Civil War

...and still unfinished until Martin Luther King Jr....and still not yet finished.

For many of us to be free, every one of us must be free. Free to make that choice between good and evil, or between a greater and a lesser good, or even between a greater or a lesser evil—with no man or woman, organization or government, working within the pretense of freedom, entitled to take away the freedom of another.

When I speak to these Chinese students, I would remind them, while giddy in their initial battle for freedom, to be careful not to lose the *war.* How often have we seen true believers wrapping themselves in the mantle of freedom, and then denying that freedom to others?

The right to use freedom to achieve one's ends implies responsibility. We have long recognized in our American society that freedom is not *license.* The father and mother might be the heads of the family, with the freedom to discipline their children, but not possessed of the freedom to beat and abuse them. A man's home may be his castle, but he has no right to burn it down and endanger others' lives or leave them without shelter. He is free to drive his automobile, but not to travel at seventy miles per hour in a thirty mile per hour zone, nor to race through a red light, nor to drive on the left-hand side of the road or while intoxicated.

I will discuss with my Chinese friends that responsible freedom suggests the need for a structure with the fine balance of a democratic society. I will gently point out the Student Rebellion in China twenty years ago, against the Gang of Four...and how that "revolution" went awry—all in the name of freedom. We must maintain a flexible arena of choice; we must understand the limits of our freedom: My freedom to swing my arms as I walk stops at the end of your nose.

Appendix

Thus, the beauty of the American brand of constitutionally validated freedom is that even in a democracy where a majority rules, no majority, no matter how large or strong, can persecute or tyrannize a minority, no matter how small or weak. Those limitations on the freedom of a majority are embedded in our Bill of Rights, in those first Ten "Unamendable" Amendments. It is undeniable that freedom is not divisible; a people cannot be half-slave and half-free. But, historically, man has always possessed two basic freedoms, and we must understand the difference. The first is natural freedom…the freedom of man on a desert island, in a state of nature, as a "noble savage." He has no restraint; he can carouse; he can lie, he can drive his car 100 miles an hour around the island, swigging bourbon. *He can hurt only himself.* But when he leaves that desert island, and joins the community of man, as Rousseau has written, he enters into social contract with his fellow man, and agrees to give up some of his natural freedoms in exchange for something else: civil freedom, where consideration must be given to the greater good of the greater number. This concession, this compromise, enables the state he has contracted with to protect him against internal and external threats, to ensure the safety of his person and his property. These stated protections may be considered as having started with the Mayflower Compact of 1620 and continue up to the Civil Rights Act of 1964.

The greatest freedom of all might well be the right to be wrong. Under this freedom, Voltaire may cry out, "I disagree with what you say, but I defend to the death your right to say it." This is the real freedom of choice, the real freedom to search out universal truth. Trial and error are both scientific and moral processes. In the long run, the will and voice of the people, we believe, will be right; but in the short run, we try and we often err in the trying. Freedom of choice will inevitably lead us to the right, and our faith in man's eventual discovery of the right has been confirmed again and again.

The balance is never perfect, but it strives for perfection. Our grasp must exceed our reach or, as Browning wrote, "What's a heaven for?" Within this delicate balance we have established minimum and maximum boundaries beyond which governmental powers cannot diminish or grow. We have established safeguards, a system of checks and balances, so that our freedom can fulfill its mission without destroying those whom it was designed to serve.

Freedom does not insure the future, but it keeps the future open. Freedom does not guarantee the good life, but it contains the seed and the promise of achieving that life. As one of our leaders, Thomas

Jefferson, wrote about a new university in 1820: "This institution will be based on the unlimited freedom of the human mind. For here we are not afraid to follow truth wherever it may lead, not to tolerate any error as long as reason is free to combat it."

There have been periods in America's quest for the unfinished business of freedom when a free man's right to examine all sides of an issue was called guilt by association; when a free man's right to criticize society was called disloyal; when a free man's right to be a nonconformist was called subversive. But free men have prevailed [sic]; they have beaten the challenge of McCarthyism, of Watergate...and we continue in our unfinished quest.

I would tell my Chinese students that there is nothing so powerful as an idea whose time has come, and it appears that the idea of freedom has caught hold in their native land. Above all, their fight for freedom must be grounded in a deep faith, in a belief in their fellow man...for, as John Faulkner wrote: "I believe in man. I believe that man is good, and I believe that man will not only survive, he will prevail."

EDITOR'S NOTE: *As we went to press, the world learned that the Chinese people were unable to escape their history of Totalitarianism and that the government has forcefully moved to suppress the movement towards democracy. This, in no way, takes away from the truths expounded in Stanley Frankel's article.*

Appendix

Clout, You Gotta Have Clout*

by Stanley A. Frankel

I joined City News a few weeks after graduating from Northwestern University in 1940, being propelled from editorship of the Daily Northwestern to the bottom CNB assignment
...night police reporter. I was warned that my summer would be not only a trial, but the hardest, meanest schooling a fledgling newspaperman could get.

My first couple of weeks at City News were a nightmare. Sent out cold to cover the South police beat from 4:00 p.m. until midnight, I would invariably be at the wrong station when big crime news broke. I felt the police were uncooperative, uncaring and contemptuous. When I politely asked the sergeant if I could go behind his desk to check the police teletype bulletins or the station blotter, I was bawled out, cursed and told to get my gabardined can out of the way.

Every night I was being scooped and didn't find out about murders or robberies on my beat until the next morning when an editor would present me with the clips from the morning paper. One hardboiled CNB copy editor began passing the word that this kid Frankel couldn't report a killing if it happened to his Thompson Restaurant waiter while being served a ham sandwich!

Searching for a solution I turned to the father of a freshman I had befriended while at Northwestern. He was Jake Arvey, Chicago's political boss. At a meeting with Arvey, I recited my troubles and that, most likely, I would be jobless in a few days. Arvey could not have been friendlier,

---

* Reprinted from *Chicago City News Bureau Centennial Book.*

recounting the many times his son had told him how I had helped at Northwestern.

Arvey picked up the phone and called Police Chief Michael Prendergast. "Mike, I'm sending a dear young friend over to see you in 15 minutes. Give him anything he wants…anything!"

I thanked him and went to Central Police Headquarters. Prendergast stood up when I walked in, put his arm around my shoulders and told me how good Arvey had been to him. "Son, whatever I can do for Jake, it's done. Tell me…how can I help you?"

I repeated my story about my problems, fingering the desk sergeants on the South Side. He said, "Son, the next time any of those bozos gives you a hard time, reach for the phone on their desk and dial 111…that's my direct line. Tell me you're calling with a problem and then give the phone to the guy at the desk. When I'm finished with him, you'll find he'll be 200 percent cooperative."

Almost floating on air, I left Prendergast's office and headed directly for the meanest adversary of all, the desk sergeant at the Wabash Av. station, to me a particularly obnoxious character who seemed to take delight in making my life miserable.

Without asking, I opened the swinging door to the cubicle, brushed aside the sergeant and helped myself to the police blotter. I dallied with the reports, then plopped the blotter down in front of him and made my way to the teletype machine, all the time feeling the sergeant's gaze. Taking notes from the teletype, I turned and faced the sergeant, whose temperament shifted from a glare to a puzzled uneasiness. "What the hell do you think you're doing?" he finally asked, but without the old bite.

"I'm doing my job as a reporter," I shot back. "You want to make something of it?" I edged toward the phone on his desk, readying myself for that magical phone call to 111.

His answer, soft and apologetic, startled me. "Hey, kid. Take it easy. Help yourself to whatever you need. Just let me make a living too."

From that moment on, in other stations and with other tough policemen, I never had a bit of trouble. Policemen became friendly and cooperative. For a while I thought Prendergast had passed the word. Not so, I learned later. There had been no instructions from headquarters. What really happened was that in the world of muscle and intimidation, to concede that you were timid was to lose. But to demonstrate you couldn't be shoved showed you knew how to handle the police in their mean and gray world.

Appendix

The Enemy Is Us, Not Saddam*

by Stanley A. Frankel

Hold on, now. The enemy is not Saddam Hussein or Arafat or Qaddafi.

We have met the enemy and he is us.

Over ten years ago, our nation, after its trauma with the OPEC oil crisis and the attendant gas station lines, put in place a Grand Plan which would have made us independent of Arab oil, certainly before 1990.

This Grand Plan included 55 miles per hour speed limits, additional exploration of continental and off-shore oil, energy conservation laws such as mandating autos using fewer gallons per mile, penalties for one-driver-percar commuters; also shifting energy use from oil to liquified [sic] natural gas, synfuels and solar, and even greater reliance on both coal and nuclear energy with a willingness to pay the extra costs for pollution control and safety.

Had this plan, and others, been fully implemented, today our young men and women would not be at risk in the Persian Gulf; nor would we give a damn what Arab tyrant ruled which emirate or sultanate or tribe.

Instead of implementation, once the crisis eased, we went back to our wasteful energy ways; we stopped enforcing conservation; and now we are compelled to repeat that lesson in recent history which we had quickly forgotten.

And we elected politicians who followed our lead, who lacked the foresight and wisdom to insist on following through on our target of energy independence. Most Americans voted for Reagan-Bush and their

---

* Reprinted from *City University Press Reporter*.

clearly outspoken pledge of laissez-faire [sic], all is well, "don't just-do something; stand there"; deregulate and don't enforce those rules.

Maybe Mencken was right: "No one ever went broke underestimating the intelligence of the American people." We shall see whether we can learn what we must now do to insure oil independence by 2000; or do we have to endure stupid, weak leadership, and not only gas lines, but body bags, again…and again.

Don't blame Reagan and Bush

…or the several Husseins…the enemy is us.

Where Have You Gone, Joe DiMaggio?*

by Stanley A. Frankel

It's July 27, Old Timers' Day at Yankee Stadium with Joe DiMaggio and the 50th anniversary of his legendary 56 game hitting streak.

This time, we were wise enough to bring along our 10-year-old grandson, Adam. Two years ago, it was just Irene and I. My wife has many interests, the least of which is baseball, but I had euchered [sic] her into accompanying me since a Yankee executive had invited us to his box and the exclusive Yankee Club restaurant before the game. We showed our special passes to the security guards, were ushered into the Club, and sat down for lunch, with our Yankee friend. Soon, an attractive young lady introduced herself to us as Jill Martin and asked if she could join us. Of course…and was she related to Billy Martin? Yes, she was his wife, which of course meant nothing to Irene.

After a pleasant, animated chat with Mrs. Martin, we ordered dessert, and Jill said: "If you'll wait here a few minutes, Billy and Mickey will be here to join us." To which one of us…she shall be nameless…inquired: "Mickey WHO?" The silence was deafening…but we quickly ate our ice cream and left.

Now…July 27…two years later…we've added Adam who has become an ardent Yankee fan, baseball-obsessed. Again, we were invited to the Yankee Club where we ordered lunch, looked around for baseball celebrities, noticed none. We were a little uneasy at the slow service and my grandson complained that he wanted to watch the oldtimers [sic] take batting practice. We skipped dessert and quickly went out to the elevator bank. About 20 fans were crowded around the elevator door…to

---

* Reprinted from *Scarsdale Inquirer*.

find that it wasn't working. With that, Adam again pouted that he was going to miss the practice session, and another gentleman, whom I was jammed up against, also complained: "I have to get down to change into my uniform."

I looked at him. It was JOE DIMAGGIO! I nudged Adam: "Look who's here!" Adam gazed up, blanched, and his eyes popped. "It's JOE!"

I turned to the baseball icon and pled: "Mr. DiMaggio, would you mind if my wife took our picture together?" He looked me in the eye, about four inches from his, and shrugged his shoulders. In a sad, sweet voice he responded: "What the hell can I do?" And as Irene snapped away, I couldn't get Adam, now shy, into the picture. But the Instamatic worked; the picture peeled off, and we got Joe to autograph the backside to Adam. Ecstasy!

But this ecstasy had at least been equalled [sic] a few months back when we took Adam to Ft. Lauderdale for two glorious weeks of Yankee spring training. Our executive friend was able to slip us into the locker room and onto the playing field, in the morning of the games, and Adam romped around the bases, had pictures taken in front of Steve Sax locker (he's Adam's #1 hero), and sat in the dugout.

The afternoons were reserved for the exhibition games, and Adam's only wish now was to meet Steve Sax, flesh to flesh. Since Sax left each game early, we went for a week before we realized he always beat us out of the locker room. Zoom. So…our Yankee friend said he'd position us on a bench near the locker room, on the way to the Yankee outside garage. When Sax left the game, we did, too, while our friend unlocked the garage gate and led us to the bench. I told Adam to stay there while I skipped to the Men's Room. Leaving the Men's Room and heading back to the bench, I noted, to my horror, that Steve Sax had left the locker room and was on his way to his car. I ran (and a 70 year old runs when his grandson's happiness is at stake), grabbed Sax around the waist as he was getting into his car, and in my most persuasive, sad voice begged: "Mr. Sax…please say hello to my grandson." He smiled

…"Sure, Pop"…and I saw

Adam, just behind Sax, waving frantically to me and calling out: "Pa…let him go. I got him already."

Sax laughed, turned to Adam, tousled his hair and asked Adam how old he was. "I'm going to be 10 tomorrow" Adam replied. What's your last name, son?" "Frankel, Mr. Sax." He took out a pen and paper and made some notes. "And where are you staying?" "At the Ft. Lauderdale

Appendix

Westin," I answered. "Well, goodbye Adam and Mr. Frankel," continued Sax as he got into his car. "Adam, you've got some grandfather."

The next morning as we awakened, I noticed an envelope had been slipped inside the hotel room door. I figured it was the bill. Opening up the envelope, I discovered it was addressed to Adam, and it was a Yankee birthday card, signed by Steve Sax and six other Yankee ballplayers!

Ecstasy, compounded! And I made a quiet wish that when I came back to this world in my next appearance, I wanted to come back as Stanley Frankel's grandson.

The birthday was a banner day...for before going to the afternoon game, we ran into Yankee left-fielder, Mel Hall, in our hotel lobby. Mel was known as a loner; he had brought his wife and daughter with him, but instead of staying with the other Yankees, he had checked into our hotel. Somewhat uneasy because of Hall's reputation as a tough customer, Adam approached him gingerly, and asked for his autograph. Hall gave him a hard look, grabbed the scorecard Adam carried, asked his name, signed the card and handed it back. "Thanks a million, Mr. Hall," glowed Adam." Hall gave us a big, broad smile and responded: "Sonny, it's an honor to be asked."

We ran to the swimming pool to show our find to Grandma who was sitting on the chaise longue with a beautiful young lady whom she introduced to us as Mrs. Hall.... Mel's wife...and to his 10 year old daughter. Adam and the daughter tossed a beachball in the pool, and as we started to leave, the little girl accosted Adam: "When my Dad gets back from the game this afternoon, he promised to play catch with me. Would you like to join us?"

To play catch with a Yankee left fielder...a 10 year old dream...yes...and a 70 year old dream...for at 5:00x p.m. that afternoon, the Halls and the Frankels tossed a rubber ball back and forth for a half hour. Adam and Mel didn't drop the ball once; the daughter missed a few; and I dropped four easy tosses!

Our anti-climax the next day was bumping into Don Mattingly, his two kids and pregnant wife outside the summer stadium, and he not only gave Adam his autograph, but with a gentle sweetness, put his arm around Adam and talked to him about school...one of his own kids was entering grade school the next fall, and his wife and he were interested in Adam's "inside" information.

All in all...a nice ending to our exquisite fortnight...which, of course, only set up the July 27 Joe DiMaggio confrontation.

Stanley A. Frankel

We had learned the answer to the plaintive Paul Simon musical question: "Where have you gone, Joe DiMaggio?" Into the Frankel lair!

"Pa, Are You Ever Going to Die?"*

by Stanley A. Frankel

Having reached that age when the older generation has disappeared and our peers are falling fast, I find it ever more difficult to sweep those sometimes harsh reminders of inevitability under the rug of my memory. My only grandson, Adam, has experienced his losses, but on another plane. For in the last year, moving from 5 to 6 years old, he has buried his pet lizard, his beloved hamster and his constant companion turtle.

Then four months ago, he was introduced to the grown-up world of permanent departure when we took him with us for the first time to a Dayton, Ohio, nursing home to visit his 89-year-old great-great uncle Arthur, my own all-time favorite relative.

Uncle Arthur brightened visibly when Adam bounced into his room, hugged his old great-great uncle and said "Let's take a walk!" To which, Uncle Arthur, promptly, but ever-so slowly, strained out of bed, dressed with the help of his nurse, asked for his cane, and then led—and was led—through the halls of Covenant House by his great-great nephew.

They were a touching sight as they slowly, painfully made their way to the therapy room, the dining room, the television room and finally into the living room, where the other 39 octogenarian patients, most in wheelchairs, were being lectured by a guest nutritionist who was pointing out that eating the right food would insure longevity. We guided Adam and Uncle Arthur to a piano, and the lecturer asked if the little boy would play something for the, until now, inattentive audience.

Adam didn't cotton to that idea at all, but I whispered to him that if he played, I would buy him an extra toy, so he sat down at the piano

---

*Reprinted from *New York Sunday Times*.

bench with Uncle Art alongside, and punched out "Chopsticks," one of the few tunes he had been taught. Uncle Art nodded to the beat of the music and the audience, coming spiritedly to life, applauded with amazing vigor for men and women major debilitating ailments and averaging 87 years old.

Adam looked to me after the applause and held up two fingers. To the group, that was both an acknowledgement of the applause as well as a sign to quiet the ovation. To me it meant: "Two toys for another tune?" and I returned the "two" sign. He punched out another quick and easy number to more applause, and then the three little fingers up in his right hand and I waved three fingers back and we went on until six tunes had been performed.

Afterward, we walked our uncle back to his room and bed, and Uncle Arthur, afflicted by the impression that he was financially destitute, reached into his pocket, pulled out his wallet and fingered a one-dollar bill. Our affluent uncle first carefully looked at the dollar, then at Adam, then at the dollar, and then again at Adam. Finally, he handed it to his little pianist with the remark: "Adam, the next time you visit me, I'm going to give you two of these." To Uncle Arthur, with his delusion of impending poverty, this was quite a generous concession.

We said our good-byes, as it happened, for the last time, left the nursing home, took a taxi to the airport and flew back. While on the plane, I told Adam that Uncle Arthur was going to be 90 years old in two months and I asked Adam if he would like to return for the birthday. Adam responded rather grimly: "Yes, I would like to go back, but, Pa, I don't think Uncle Arthur is going to make it."

And he didn't. He passed away two weeks later. We returned for the funeral without Adam, but then upon our return home told him the bad news. His eyes teared up, but he insisted on all the details, probably measuring this event against the deaths of his pets: "Did you bury Uncle Arthur? Did you dig the hole? How big was the hole? Did you bury him in a box?"

"Yes, Adam, I supervised the digging of the hole, and the box was a beautiful wooden box called a casket, and flowers were piled on the casket and it was lowered into the grave, and then earth was put on top of the casket."

Adam was tearing up again at these grim details, so to console him, I explained: "Adam, in the years to come, a long, long time from now, we'll all be with Uncle Arthur again."

And Adam, his tunnel vision focusing only on the black hole and the dirt on top of the casket, responded bluntly and unequivocally, "Uncle Arthur is gone forever."

Then he turned sorrowfully to me, his grandfather, "Pa," who had that special relationship with a grandson which only you grandads understand. He asked: "Pa, are you ever going to die?"

"Oh, Adam, such a question. We all finally die, but it will be years and years from now."

To which Adam responded by grabbing onto me, burying his head in my shoulder, and sobbing uncontrollably.

To which, I hugged him back—and cried a little, too.

Stanley A. Frankel

Appendix

Angry with Senator Quayle, Soldier?

by Stanley Frankel

I think I have made clear in the introduction to this book, and perhaps more subtly, throughout, my aversion to being shot at by—and shooting at—fellow human beings. These feelings rose to the surface when I read in 1988 that the then-vice-presidential candidate and current vice-president Dan Quayle had tried hard to get out of going to Viet Nam, mainly by applying for a post in the Indiana National Guard which did not appear headed to Nam. So, I wrote this tongue-in-cheek column about his successful efforts—and my own failed attempts—to stay as far away from combat zones as possible.

This World War II combat veteran isn't mad at Senator Quayle for his success in avoiding the horrors of war. Rather, I am full of admiration and sorry I didn't have some of his smarts and connections. My own efforts to beat the draft and, then, infantry-action are redolent with stupidity and blunders.

When the draft was legislated in 1940, I was working in New York. I opted (cleverly I thought) to register out of my hometown, Dayton Ohio, rather than New York City because my favorite Uncle Max was chairman of the Dayton draft board. When the draft lottery was held, sure enough, I won first place and soon received instructions to report for my draft physical exam. Not to worry.

I phoned Uncle Max, who, I was sure, wouldn't let me be dragged into the army. Wrong. He congratulated me enthusiastically, opined that a one year army hitch would be good for me and pointedly advised me not to be late for my physical.

245

But I did have another ace: my eyes were 20/400, without glasses, far below minimum army requirements. I was, again, so sure the army medicos would not accept me that I told my New York roommate not to pack up my things. I'd be back from Dayton in a few days.

The army oculist asked me to read the top lines on the chart, and I responded with the hoary gag, which was no joke at all to me: "Doc, I can't even see the chart." He laughed and patted me on the back and guessed that the army could always find some job for a near-blind draftee. After all, it was only for 12 months. He okayed my eye exam.

I entrained for Camp Shelby, Miss. to join the Ohio 37th National Guard Division. I was assigned to the finance department, a fairly safe haven.

More disaster, however: draft enlistment was extended in October for another year; and in December, Pearl Harbor. Two months later, my division boarded the SS President Coolidge, bound for Fiji.

Suva, in Fiji wasn't bad duty. For six months, my work consisted of paying the troops. Then, my finance colonel suggested I attend the Jungle Warfare Officers School being run by returning Guadalcanal officers. He promised that if I passed the 90-day OCS, he'd pull me back to Finance where I would serve out the war in the rear echelons of the division, but as an officer not an enlisted man. I ranked last among the 130 candidates in weaponry, agility, short-order drill and foxhole digging.

On graduation day, the newly commissioned second lieutenants assembled in the mess hall to learn their next assignments. The commandant
came to my name: "Frankel, assigned to Co. F., 148th Infantry Regiment." What a horrible mistake. I dashed up to the commandant and told him of his error. "No mistake, Frankel. You were requested by Finance but there is a terrible shortage of platoon leaders, and we've decided anyone who graduates from this course will go into the infantry."

"Sir, may I resign my commission?"

"You forget it or I'll have you court-martialed."

So, this near-blind, bumbling new officer soon became the worst platoon leader in the South Pacific theater. When they asked me what kind of weapon I wanted to take into combat, I told the supply sgt: "A seeing eye dog."

One month later, we hit New Georgia, Solomon Islands, up the chain from Guadalcanal; and the old adage was confirmed: "Good platoon leaders get killed and bad ones get their men killed." After a month of mud and blood and dysentery, we had won the ugly little island and my platoon had been decimated.

Up the Solomon ladder we went

...next to Bougainville...followed by another request for transfer, a request denied for the same reasons

...then Manila in the Phillippines [sic]

...request submitted and denied...a mountain flight for Baguio...more consolation prizes, but no transfer.

It had become apparent that I led a charmed life. Men were being hit all around me, and I hadn't been grazed. The rumor mill in the regiment ground out the real poop on Captain Frankel's invulnerability. The Japanese had made a tactical decision not to shoot at Frankel...he was their best chance to win the war.

And finally...not one...not two...not three...not four years after my induction...but five years later...while we were chasing Japanese up the Cagayan Valley, the Japanese surrendered. I had amassed so many points awarded for years overseas, months in combat, medals accumulated for surviving, that I figured I would be sent home at once.

Another disaster: I received a summons from Commanding General Robert S. Beightler. In his tent, he returned my salute with a warm handshake and a generous comment: "Major Frankel, I have been aware of your interest in writing, and I felt bad having to turn down your many requests to be transferred to a writing assignment. However, you know there was a shortage of qualified infantry officers. Now I am happy to tell you that I have decided to honor your request."

"But General, the war is over. I didn't make any request. I want to go home."

"Major, someone has to stay here with my staff and me for the next few months to assemble and draft the history of division, and we have chosen you. After all, Major, you realize that there is a peace to be won."

We finally sailed back to San Francisco, received our discharges, and went back home. My uncle Max was on hand to greet me. Four and a half years beyond the time he had promised. But I decided against bitterness and recrimination. I had quietly pledged to myself during those bad years that if I got out of this period alive and in one piece, I would never complain about the army, or anything else, again.

And so, I am not really complaining here. I admire and envy the Republican vice-presidential candidate for pulling off, with ease, what I had tried and failed to do many times in many different ways.

Mad at him? I quail at the thought.

Stanley A. Frankel

Frankel-y Speaking[*]

by Stanley Frankel

Fittingly, and last, a tribute I wrote in college to my closest friend and cousin, Jack "Pee-Wee" Margolis, who died our senior year at age 21. This article is presented in memory of all of my beloved relatives and friends who have preceded me off the stage: my sainted mother and my dad (who died at age 34); my sister, Phyllis, who never had a chance; my aunts and uncles who surrogated my childhood: Dot and Bart, Arthur and Ruth, Max and Roma, Janet, Ethel and Joe, Fanny and Jacob; first cousin and first friend Jack Coney; friends Norman Gitman, Dotty and Frank Weprin, Alvin Ablon, Rudy Van Dyke, Bordy Greathouse, Phyllis Kessel Finn, Persis Gladieux, Hubert Humphrey, Adlai Stevenson, and Bobby Kennedy, Irene's folks: Bessie and Salem Baskin and Uncle Neel. I pray they all know that they will never be forgotten "in war and peace," "from here to eternity."

Jack always liked to fool around with chemicals; with me, it was a typewriter. Everyone said that he had sulfuric acid running through his veins—and I had printer's ink.

For Jack, you see, was my best friend—a senior and honor student at Purdue. And I said my last goodbye to Jack one week ago at a little cemetery in Dayton, Ohio.

Jack died of an incurable blood disease. He was a victim of certain chemical reactions of the blood, dread chemical reactions which he himself had often planned to eradicate.

---

[*] Reprinted from *Daily Northwestern*

And now I can't forget the twentyone [sic] years we spent together. The struggle for grades and the football games and the double dates, everything comes back to me clearly. And through our close association, a certain theme was prevalent.

That theme was our plans for the future. Jack always felt deeply the suffering of others. This sensitiveness to the feelings of his fellowmen was one of his greatest characteristics, and many times as we talked together, he revealed to me his overwhelming passion to become a great chemist so that he could contribute something to eliminate the disease and the pain of others.

We dreamed our dreams for twenty-one years. His highest goal probably lay in discovering a cure for cancer, anonymously. Mine, I must admit, lay in fields bordering, more closely, the heroic and the glorious.

Yet his heart and his dreams were both bigger than his frail body. And when he came home from Purdue several weeks ago, he was suffering from this deadly blood disease.

All of us pitched in to help. I gave him a blood tranfusion [sic] when he was very low. Wonder of wonders, it looked for awhile as if he would be the one in one thousand to survive.

And last week, just two days before he died, the doctors were quite optimistic, and they claimed that he would recover, barring unforeseen complications.

But the very night of the optimistic hopes, he began to fade away. Queer things—chemical reactions—were happening to his body.

As he lay on his deathbed, unable to lift even a finger and yet clear of mind, he kept repeating in prayer, "Oh God, give me a break!"

"Give me a break," he asked. A mind which gave promise of so many contributions to human good still nurtured hopes for the future. He wanted to live—for many reasons to be sure—but one of the strongest motives was his own desire to utilize his education and his wisdom and his ingrained ingenuity so that others like himself might live.

And as he pleaded with his Maker to save him so that he might save others, those queer chemical reactions with which man cannot cope kept on happening. And he died of leukemia—another victim of an incurable blood disease against which science must accede to prayer.

But, as he died, still pleading for life, his case reached the ears of specialists in this region. And upon his death, a complete examination of his body by competent medical authorities gave those authorities the answer to many of the questions on chemical reaction which, until this time, had remained unsolved. These answers are of such universal interest that they will soon be published in a medical journal, and the doctors feel that a step in the direction of overcoming this disease has been taken.

And so Jack never knew that, in death, he perhaps accomplished what would have taken him many more years of life to achieve. He was overcome by chemical reactions which he had hoped to understand and to prevent. And in this tragic end, lay perhaps the realization of his hopes and his dreams.

Ironically enough, Jack would have given a lifetime to find the cause of and the cure for leukemia. He might have done just that in a glorious lifetime of twenty-one years.

www.ingramcontent.com/pod-product-compliance
Lightning Source LLC
Chambersburg PA
CBHW051647040426
42446CB00009B/1021